陈恒 朱毅 项聪 编著

JSP
教学做一体化教程
网站设计

清华大学出版社
北京

内 容 简 介

JSP技术是基于Java语言的一种动态交互式网页技术标准,它由Sun公司倡导,多家公司共同参与制定,并于1999年由Sun公司正式公布。本书采用"教学做"一体化的方式撰写,合理地组织学习单元,并将每个单元分解为核心知识、能力目标、任务驱动、实践环节四个模块。全书共分10章,内容包括JSP简介及开发环境的构建、JSP语法、JSP内置对象、JSP与JavaBean、JSP访问数据库、Java Servlet基础、基于Servlet的MVC模式、过滤器、EL与JSTL、地址簿管理信息系统等重要内容。书中实例侧重实用性和启发性,趣味性强、通俗易懂,使读者能够快速掌握JSP网站设计的基础知识与编程技巧,为适应实战应用打下坚实的基础。

本书适合作为高等院校相关专业的教、学、做一体化教材,也适合作为JSP网站设计培训教材,还可以作为JSP网站设计爱好的自学读物。

本书封面贴有清华大学出版社防伪标签,无标签者不得销售。
版权所有,侵权必究。侵权举报电话:010-62782989 13701121933

图书在版编目(CIP)数据

JSP网站设计教学做一体化教程/陈恒,朱毅,项聪编著. —北京:清华大学出版社,2012.11(2017.3重印)
ISBN 978-7-302-29889-2

Ⅰ.①J… Ⅱ.①陈… ②朱… ③项… Ⅲ.①JAVA语言－网页制作工具－程序设计－教材 Ⅳ.①TP312 ②TP393.092

中国版本图书馆CIP数据核字(2012)第198660号

责任编辑:田在儒
封面设计:李 丹
责任校对:袁 芳
责任印制:沈 露

出版发行:清华大学出版社
网　　址:http://www.tup.com.cn, http://www.wqbook.com
地　　址:北京清华大学学研大厦A座　　　　邮　编:100084
社 总 机:010-62770175　　　　　　　　　　邮　购:010-62786544
投稿与读者服务:010-62776969, c-service@tup.tsinghua.edu.cn
质量反馈:010-62772015, zhiliang@tup.tsinghua.edu.cn
印 装 者:北京九州迅驰传媒文化有限公司
经　　销:全国新华书店
开　　本:185mm×260mm　　印　张:14.5　　字　数:350千字
版　　次:2012年11月第1版　　　　　　　　印　次:2017年3月第4次印刷
印　　数:4301~4500
定　　价:29.00元

产品编号:045655-01

本教材按照教、学、做一体化模式精编了 JSP 的核心内容,以核心知识、能力目标、任务驱动和实践环节为单元组织本教材的体系结构。核心知识体现最重要和实用的知识,是教师需要重点讲解的内容;能力目标提出学习核心知识后应具备的编程能力;任务驱动给出了教师和学生共同来完成的任务;实践环节给出了需要学生独立完成的实践活动。

全书共分 10 章。第 1 章主要介绍了 JSP 运行环境的构建,并通过一个简单的 Web 应用介绍了 JSP 项目开发的基本步骤。第 2 章讲述了 JSP 语法,包括 Java 脚本元素以及常用的 JSP 标记。第 3 章介绍了常见的 JSP 内置对象,包括 request、response、session、application 以及 out。第 4 章介绍了 JSP 与 JavaBean,JSP 和 JavaBean 技术的结合不仅可以实现数据的表示和处理分离,而且可以提高 JSP 程序代码重用的程度,是 JSP 编程中常用的技术。第 5 章详细地介绍了在 JSP 中如何访问关系数据库,如 Oracle、SQL Server、MySQL 和 Microsoft Access 等数据库。第 6 章、第 7 章讲述了 Servlet 的运行原理以及基于 Servlet 的 MVC 模式。第 8 章详细地讲述了过滤器的概念、运行原理以及实际应用。过滤器可以过滤浏览器对服务器的请求,也可以过滤服务器对浏览器的响应。第 9 章主要介绍了 EL 与 JSTL 核心标签库的基本用法。第 10 章通过一个综合案例讲述了如何采用 JSP+JavaBean+Servlet 的模式来开发一个 Web 应用。

本教材特别注重引导学生参与课堂教学活动,适合高等院校相关专业作为教、学、做一体化的教材。

编 者
2012 年 10 月

第 1 章　JSP 简介及开发环境的构建 ……………………………………………… 1

  1.1　构建开发环境 ………………………………………………………………… 1
      1.1.1　核心知识 …………………………………………………………… 1
      1.1.2　能力目标 …………………………………………………………… 2
      1.1.3　任务驱动 …………………………………………………………… 2
      1.1.4　实践环节 …………………………………………………………… 7
  1.2　使用 Eclipse 开发 Web 应用 ………………………………………………… 7
      1.2.1　核心知识 …………………………………………………………… 7
      1.2.2　能力目标 …………………………………………………………… 7
      1.2.3　任务驱动 …………………………………………………………… 7
      1.2.4　实践环节 …………………………………………………………… 11
  1.3　小结 …………………………………………………………………………… 11
  习题 1 ……………………………………………………………………………… 12

第 2 章　JSP 语法 ……………………………………………………………………… 13

  2.1　JSP 页面的基本构成 ………………………………………………………… 13
      2.1.1　核心知识 …………………………………………………………… 13
      2.1.2　能力目标 …………………………………………………………… 13
      2.1.3　任务驱动 …………………………………………………………… 13
      2.1.4　实践环节 …………………………………………………………… 15
  2.2　Java 程序片 …………………………………………………………………… 15
      2.2.1　核心知识 …………………………………………………………… 15
      2.2.2　能力目标 …………………………………………………………… 16
      2.2.3　任务驱动 …………………………………………………………… 16
      2.2.4　实践环节 …………………………………………………………… 18
  2.3　成员变量和方法的定义 ……………………………………………………… 18
      2.3.1　核心知识 …………………………………………………………… 18
      2.3.2　能力目标 …………………………………………………………… 18
      2.3.3　任务驱动 …………………………………………………………… 19

  2.3.4 实践环节 ·· 20

2.4 Java 表达式 ·· 20

  2.4.1 核心知识 ·· 20

  2.4.2 能力目标 ·· 20

  2.4.3 任务驱动 ·· 20

  2.4.4 实践环节 ·· 21

2.5 page 指令标记 ·· 21

  2.5.1 核心知识 ·· 22

  2.5.2 能力目标 ·· 22

  2.5.3 任务驱动 ·· 22

  2.5.4 实践环节 ·· 23

2.6 include 指令标记 ·· 23

  2.6.1 核心知识 ·· 23

  2.6.2 能力目标 ·· 24

  2.6.3 任务驱动 ·· 24

  2.6.4 实践环节 ·· 25

2.7 include 动作标记 ·· 25

  2.7.1 核心知识 ·· 25

  2.7.2 能力目标 ·· 26

  2.7.3 任务驱动 ·· 26

  2.7.4 实践环节 ·· 27

2.8 forward 动作标记 ··· 27

  2.8.1 核心知识 ·· 27

  2.8.2 能力目标 ·· 27

  2.8.3 任务驱动 ·· 27

  2.8.4 实践环节 ·· 29

2.9 param 动作标记 ··· 29

  2.9.1 核心知识 ·· 29

  2.9.2 能力目标 ·· 29

  2.9.3 任务驱动 ·· 30

  2.9.4 实践环节 ·· 31

2.10 小结 ·· 31

习题 2 ··· 32

第 3 章 JSP 内置对象 ··· 33

3.1 请求对象 request ·· 33

  3.1.1 核心知识 ·· 33

  3.1.2 能力目标 ·· 34

  3.1.3 任务驱动 ·· 34

3.1.4　实践环节 …………………………………………………………… 37
　3.2　响应对象 response ……………………………………………………………… 37
　　　3.2.1　核心知识 …………………………………………………………… 37
　　　3.2.2　能力目标 …………………………………………………………… 38
　　　3.2.3　任务驱动 …………………………………………………………… 38
　　　3.2.4　实践环节 …………………………………………………………… 42
　3.3　会话对象 session ………………………………………………………………… 42
　　　3.3.1　核心知识 …………………………………………………………… 42
　　　3.3.2　能力目标 …………………………………………………………… 43
　　　3.3.3　任务驱动 …………………………………………………………… 43
　　　3.3.4　实践环节 …………………………………………………………… 51
　3.4　全局应用程序对象 application ………………………………………………… 51
　　　3.4.1　核心知识 …………………………………………………………… 51
　　　3.4.2　能力目标 …………………………………………………………… 52
　　　3.4.3　任务驱动 …………………………………………………………… 52
　　　3.4.4　实践环节 …………………………………………………………… 54
　3.5　小结 …………………………………………………………………………… 54
　习题 3 …………………………………………………………………………………… 54

第 4 章　JSP 与 JavaBean …………………………………………………………… 56

　4.1　编写 JavaBean ………………………………………………………………… 56
　　　4.1.1　核心知识 …………………………………………………………… 56
　　　4.1.2　能力目标 …………………………………………………………… 57
　　　4.1.3　任务驱动 …………………………………………………………… 57
　　　4.1.4　实践环节 …………………………………………………………… 58
　4.2　JSP 页面中创建与使用 bean …………………………………………………… 58
　　　4.2.1　核心知识 …………………………………………………………… 58
　　　4.2.2　能力目标 …………………………………………………………… 58
　　　4.2.3　任务驱动 …………………………………………………………… 59
　　　4.2.4　实践环节 …………………………………………………………… 61
　4.3　获取 bean 的属性 ……………………………………………………………… 61
　　　4.3.1　核心知识 …………………………………………………………… 61
　　　4.3.2　能力目标 …………………………………………………………… 61
　　　4.3.3　任务驱动 …………………………………………………………… 62
　　　4.3.4　实践环节 …………………………………………………………… 63
　4.4　修改 bean 的属性 ……………………………………………………………… 64
　　　4.4.1　核心知识 …………………………………………………………… 64
　　　4.4.2　能力目标 …………………………………………………………… 64
　　　4.4.3　任务驱动 …………………………………………………………… 64

    4.4.4 实践环节 ··· 67
  4.5 JSP 与 bean 结合的简单例子 ··· 67
    4.5.1 核心知识 ··· 67
    4.5.2 能力目标 ··· 67
    4.5.3 任务驱动 ··· 68
    4.5.4 实践环节 ··· 70
  4.6 小结 ··· 70
 习题 4 ··· 70

第 5 章　JSP 访问数据库 ··· 73

  5.1 使用 JDBC-ODBC 桥接器连接数据库 ····································· 73
    5.1.1 核心知识 ··· 73
    5.1.2 能力目标 ··· 74
    5.1.3 任务驱动 ··· 74
    5.1.4 实践环节 ··· 78
  5.2 使用纯 Java 数据库驱动程序连接数据库 ································· 78
    5.2.1 核心知识 ··· 78
    5.2.2 能力目标 ··· 79
    5.2.3 任务驱动 ··· 79
    5.2.4 实践环节 ··· 82
  5.3 Statement、ResultSet 的使用 ··· 82
    5.3.1 核心知识 ··· 82
    5.3.2 能力目标 ··· 82
    5.3.3 任务驱动 ··· 83
    5.3.4 实践环节 ··· 87
  5.4 游动查询 ··· 88
    5.4.1 核心知识 ··· 88
    5.4.2 能力目标 ··· 88
    5.4.3 任务驱动 ··· 88
    5.4.4 实践环节 ··· 91
  5.5 访问 Excel 电子表格 ··· 91
    5.5.1 核心知识 ··· 91
    5.5.2 能力目标 ··· 91
    5.5.3 任务驱动 ··· 92
    5.5.4 实践环节 ··· 93
  5.6 使用连接池 ··· 94
    5.6.1 核心知识 ··· 94
    5.6.2 能力目标 ··· 94
    5.6.3 任务驱动 ··· 94

　　　　5.6.4　实践环节 ………………………………………………………… 97
　5.7　其他典型数据库的连接 ………………………………………………… 97
　　　　5.7.1　核心知识 …………………………………………………………… 97
　　　　5.7.2　能力目标 …………………………………………………………… 98
　　　　5.7.3　任务驱动 …………………………………………………………… 98
　　　　5.7.4　实践环节 ………………………………………………………… 101
　5.8　PreparedStatement 的使用 …………………………………………… 101
　　　　5.8.1　核心知识 ………………………………………………………… 101
　　　　5.8.2　能力目标 ………………………………………………………… 102
　　　　5.8.3　任务驱动 ………………………………………………………… 102
　　　　5.8.4　实践环节 ………………………………………………………… 107
　5.9　小结 …………………………………………………………………… 107
　习题 5 ………………………………………………………………………… 107

第 6 章　Java Servlet 基础 …………………………………………………… 109

　6.1　Servlet 类与 servlet 对象 ……………………………………………… 109
　　　　6.1.1　核心知识 ………………………………………………………… 109
　　　　6.1.2　能力目标 ………………………………………………………… 109
　　　　6.1.3　任务驱动 ………………………………………………………… 109
　　　　6.1.4　实践环节 ………………………………………………………… 111
　6.2　servlet 对象的创建与运行 …………………………………………… 111
　　　　6.2.1　核心知识 ………………………………………………………… 111
　　　　6.2.2　能力目标 ………………………………………………………… 111
　　　　6.2.3　任务驱动 ………………………………………………………… 111
　　　　6.2.4　实践环节 ………………………………………………………… 113
　6.3　通过 JSP 页面访问 servlet …………………………………………… 114
　　　　6.3.1　核心知识 ………………………………………………………… 114
　　　　6.3.2　能力目标 ………………………………………………………… 114
　　　　6.3.3　任务驱动 ………………………………………………………… 114
　　　　6.3.4　实践环节 ………………………………………………………… 116
　6.4　doGet 和 doPost 方法 ………………………………………………… 116
　　　　6.4.1　核心知识 ………………………………………………………… 117
　　　　6.4.2　能力目标 ………………………………………………………… 117
　　　　6.4.3　任务驱动 ………………………………………………………… 117
　　　　6.4.4　实践环节 ………………………………………………………… 119
　6.5　重定向与转发 …………………………………………………………… 120
　　　　6.5.1　核心知识 ………………………………………………………… 120
　　　　6.5.2　能力目标 ………………………………………………………… 120
　　　　6.5.3　任务驱动 ………………………………………………………… 120

　　　　　　6.5.4　实践环节 ………………………………………………………… 123
　6.6　在 servlet 中使用 session ……………………………………………………… 123
　　　　　　6.6.1　核心知识 ………………………………………………………… 123
　　　　　　6.6.2　能力目标 ………………………………………………………… 123
　　　　　　6.6.3　任务驱动 ………………………………………………………… 124
　　　　　　6.6.4　实践环节 ………………………………………………………… 126
　6.7　小结 …………………………………………………………………………… 126
　习题 6 ……………………………………………………………………………… 127

第 7 章　基于 Servlet 的 MVC 模式 ………………………………………………… 128

　7.1　JSP 中的 MVC 模式 …………………………………………………………… 128
　　　　　　7.1.1　核心知识 ………………………………………………………… 128
　　　　　　7.1.2　能力目标 ………………………………………………………… 129
　　　　　　7.1.3　任务驱动 ………………………………………………………… 129
　　　　　　7.1.4　实践环节 ………………………………………………………… 134
　7.2　使用 MVC 模式查询数据库 …………………………………………………… 134
　　　　　　7.2.1　核心知识 ………………………………………………………… 134
　　　　　　7.2.2　能力目标 ………………………………………………………… 134
　　　　　　7.2.3　任务驱动 ………………………………………………………… 135
　　　　　　7.2.4　实践环节 ………………………………………………………… 141
　7.3　小结 …………………………………………………………………………… 141
　习题 7 ……………………………………………………………………………… 142

第 8 章　过滤器 ……………………………………………………………………… 143

　8.1　Filter 类与 filter 对象 …………………………………………………………… 143
　　　　　　8.1.1　核心知识 ………………………………………………………… 143
　　　　　　8.1.2　能力目标 ………………………………………………………… 143
　　　　　　8.1.3　任务驱动 ………………………………………………………… 144
　　　　　　8.1.4　实践环节 ………………………………………………………… 145
　8.2　filter 对象的部署与运行 ……………………………………………………… 145
　　　　　　8.2.1　核心知识 ………………………………………………………… 145
　　　　　　8.2.2　能力目标 ………………………………………………………… 145
　　　　　　8.2.3　任务驱动 ………………………………………………………… 145
　　　　　　8.2.4　实践环节 ………………………………………………………… 147
　8.3　登录验证过滤器的实现 ………………………………………………………… 147
　　　　　　8.3.1　核心知识 ………………………………………………………… 147
　　　　　　8.3.2　能力目标 ………………………………………………………… 147
　　　　　　8.3.3　任务驱动 ………………………………………………………… 147
　　　　　　8.3.4　实践环节 ………………………………………………………… 151

8.4 小结 ……………………………………………………………………… 151
习题 8 ………………………………………………………………………… 152

第 9 章　EL 与 JSTL ……………………………………………………… 153

9.1 使用 EL 访问对象的属性 ……………………………………………… 153
　　9.1.1 核心知识 …………………………………………………… 153
　　9.1.2 能力目标 …………………………………………………… 154
　　9.1.3 任务驱动 …………………………………………………… 154
　　9.1.4 实践环节 …………………………………………………… 156
9.2 EL 内置对象 …………………………………………………………… 156
　　9.2.1 核心知识 …………………………………………………… 156
　　9.2.2 能力目标 …………………………………………………… 158
　　9.2.3 任务驱动 …………………………………………………… 158
　　9.2.4 实践环节 …………………………………………………… 159
9.3 基本输入输出标签 ……………………………………………………… 160
　　9.3.1 核心知识 …………………………………………………… 160
　　9.3.2 能力目标 …………………………………………………… 161
　　9.3.3 任务驱动 …………………………………………………… 161
　　9.3.4 实践环节 …………………………………………………… 162
9.4 流程控制标签 …………………………………………………………… 163
　　9.4.1 核心知识 …………………………………………………… 163
　　9.4.2 能力目标 …………………………………………………… 163
　　9.4.3 任务驱动 …………………………………………………… 163
　　9.4.4 实践环节 …………………………………………………… 165
9.5 迭代标签 ………………………………………………………………… 165
　　9.5.1 核心知识 …………………………………………………… 165
　　9.5.2 能力目标 …………………………………………………… 166
　　9.5.3 任务驱动 …………………………………………………… 166
　　9.5.4 实践环节 …………………………………………………… 167
9.6 小结 ……………………………………………………………………… 167
习题 9 ………………………………………………………………………… 168

第 10 章　地址簿管理信息系统 …………………………………………… 169

10.1 系统设计 ……………………………………………………………… 169
　　10.1.1 系统功能需求 …………………………………………… 169
　　10.1.2 系统模块划分 …………………………………………… 169
10.2 数据库设计 …………………………………………………………… 170
　　10.2.1 数据库概念结构设计 …………………………………… 170
　　10.2.2 数据库逻辑结构设计 …………………………………… 170

        10.2.3 创建数据表 …… 171
    10.3 系统管理 …… 172
        10.3.1 导入相关的 jar 包 …… 172
        10.3.2 JSP 页面管理 …… 172
        10.3.3 组件与 servlet 管理 …… 177
        10.3.4 配置文件管理 …… 177
    10.4 组件设计 …… 180
        10.4.1 过滤器 …… 180
        10.4.2 数据库连接与关闭 …… 182
        10.4.3 实体模型 …… 184
        10.4.4 业务模型 …… 185
    10.5 系统实现 …… 191
        10.5.1 用户注册 …… 191
        10.5.2 用户登录 …… 195
        10.5.3 添加朋友信息 …… 197
        10.5.4 查询朋友信息 …… 202
        10.5.5 修改朋友信息 …… 205
        10.5.6 删除朋友信息 …… 213
        10.5.7 修改密码 …… 215
        10.5.8 退出系统 …… 218

第1章 JSP 简介及开发环境的构建

本章主要内容

- 开发环境的构建
- 使用 Eclipse 开发 Web 应用

JSP 是 Java Server Pages(Java 服务器页面)的简称,是基于 Java 语言的一种 Web 应用开发技术,由 Sun 公司倡导,多家公司共同参与建立的一种动态网页技术标准。

在学习 JSP 之前,读者应具有 Java 语言基础以及 HTML 语言方面的知识。本章 1.2 节通过一个简单的 Web 应用介绍了 JSP 项目开发的基本步骤,这些基本步骤对后续章节的学习是极其重要的。

1.1 构建开发环境

1.1.1 核心知识

所谓"工欲善其事,必先利其器",在开发 JSP 应用程序前,需要构建其开发环境。

1. Java 开发工具包(JDK)

JSP 引擎需要 Java 语言的核心库和相应编译器,在安装 JSP 引擎之前,需要安装 Java 标准版(Java SE)提供的开发工具包 JDK。登录 http://www.oracle.com/technetwork/java,在 Software Downloads 里选择 Java SE 提供的 JDK,比如 Java Platform (JDK) 7u3。对于 Windows 操作系统,选择下载 jdk-7u3-windows-i586.exe,如果使用 64 位机或其他的操作系统,可以在下载列表中选择下载相应的 JDK。

2. JSP 引擎

运行包含 JSP 页面的 Web 项目还需要一个支持 JSP 的 Web 服务软件,该软件也称做 JSP 引擎或 JSP 容器,通常将安装了 JSP 引擎的计算机称为一个支持 JSP 的 Web 服务器。目前,比较常用的 JSP 引擎包括 Tomcat、JRun、Resin 等,本书采用的是 Tomcat。

登录 Apache 软件基金会的官方网站 http://jakarta.Apache.org/tomcat,下载 Tomcat 6.0 的免安装版本(即:ZIP 文件 apache-tomcat-6.0.35.zip)。登录网站后,首先在 Download 里选择 Tomcat 6.0,然后在 Binary Distributions 的 Core 中选择 zip 即可。

3. Eclipse

为了提高开发效率，通常还需要安装 IDE（集成开发环境）工具，在本书中使用的 IDE 工具是 Eclipse。Eclipse 是一个可用于开发 JSP 程序的 IDE 工具。登录 http://www.eclipse.org，在 Downloads 里选择 Eclipse IDE for Java EE Developers，然后在 Download Links 里选择适用 Windows 32-bit 的 Eclipse。如果读者使用 64 位机或其他的操作系统，可以在下载列表中选择下载相应的 Eclipse。

1.1.2 能力目标

安装与配置 JSP 的运行环境。

1.1.3 任务驱动

任务的主要内容如下。

1. JDK 的安装与配置

1）安装 JDK

双击下载后的 jdk-7u3-windows-i586.exe 文件图标出现安装向导界面，选择接受软件安装协议。建议采用默认的安装路径 C:\Program Files\Java\jdk1.7.0_03。需要注意的是，在安装 JDK 的过程中，JDK 还额外提供一个 Java 运行环境 JRE（Java Runtime Environment），并提示是否修改 JRE 默认的安装路径 C:\Program Files\Java\JRE7，建议采用该默认的安装路径。

2）配置系统环境变量

安装 JDK 平台之后需要进行几个系统环境变量的设置。

（1）配置系统环境变量 Java_Home

右击桌面"我的电脑"图标，在弹出的菜单中选择"属性"命令，弹出"系统属性"对话框，选择名为"高级"的选项卡，单击"环境变量"按钮，在"新建系统变量"对话框的"变量名"文本框中输入：

```
Java_Home
```

在"变量值"文本框中输入：

```
C:\Program Files\Java\jdk1.7.0_03
```

如图 1.1 所示。

（2）编辑系统环境变量 Path

双击系统环境变量 Path 进行编辑操作，在"变量值"文本框中输入：

```
C:\Program Files\Java\jdk1.7.0_03\bin;
```

如图 1.2 所示。

（3）配置系统环境变量 classpath

在"新建系统变量"对话框的"变量名"文本框中输入：

图 1.1　新建 Java_Home 　　　　　　图 1.2　编辑 Path

classpath

在"变量值"文本框中输入：

C:\Program Files\Java\jdk1.7.0_03\jre\lib\rt.jar;.;

如图 1.3 所示。

2. Tomcat 的安装与启动

安装 Tomcat 之前需要事先安装 JDK。将下载的 apache-tomcat-6.0.35.zip 解压到磁盘的某个分区中，比如解压到 E:\，解压缩后将出现如图 1.4 所示的目录结构。执行 Tomcat 根目录 bin 文件夹中的 startup.bat 来启动 Tomcat 服务器。执行 startup.bat 启动 Tomcat 服务器会占用一个 MS-DOS 窗口，出现如图 1.5 所示的界面，如果关闭当前 MS-DOS 窗口将关闭 Tomcat 服务器。

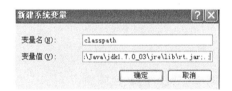

图 1.3　新建 classpath　　　　　　图 1.4　Tomcat 服务器目录结构

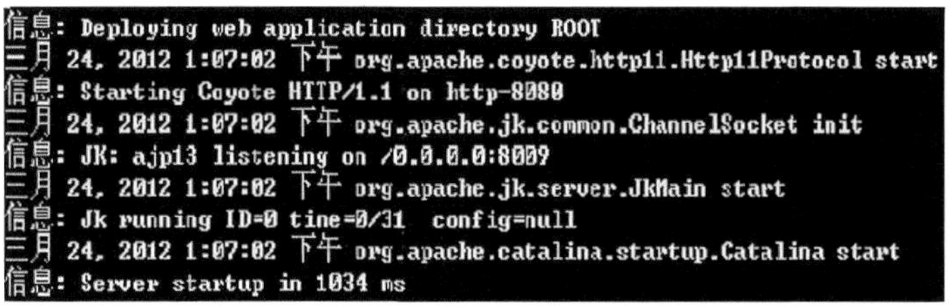

图 1.5　执行 startup.bat 启动 Tomcat 服务器

Tomcat 服务器启动后，在浏览器的地址栏中输入 http://localhost:8080，将出现如图 1.6 所示的 Tomcat 测试页面。

3. Eclipse 的安装与配置

使用 Eclipse 开发 JSP 程序之前，需要对 JDK、Tomcat 和 Eclipse 进行一些必要的配置。因此，在安装 Eclipse 之前，应该事先安装 JDK 和 Tomcat。

1）安装 Eclipse

Eclipse 下载完成后，解压到自己设置的路径下，即可完成安装。Eclipse 安装完成后就

图 1.6　Tomcat 测试页面

可以启动了。双击 Eclipse 安装目录下的 eclipse.exe 文件,启动 Eclipse。在初次启动时,需要设置工作空间,比如将工作空间设置为 E:\eclipse\workspace,如图 1.7 所示。

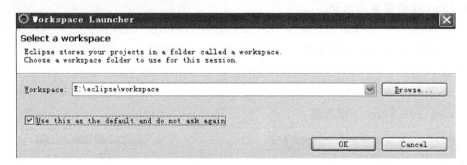

图 1.7　设置工作空间

在图 1.7 所示的对话框中单击 OK 按钮进入如图 1.8 所示的欢迎画面。

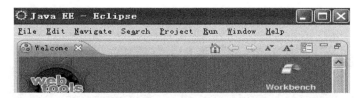

图 1.8　Eclipse 欢迎画面

2)配置 Eclipse

(1)配置 Tomcat。启动 Eclipse,选择 Window→Preferences 菜单项,在弹出的对话框中选择 Server→Runtime Environments 选项,如图 1.9 所示。

(2)单击 Add 按钮后,弹出如图 1.10 所示的 New Server Runtime Environment 窗口,在此可以配置各种版本的 Web 服务器。

(3)选择 Apache Tomcat v6.0 服务器版本,单击 Next 按钮,进入如图 1.11 所示的窗口。

(4)单击 Browse 按钮,选择 Tomcat 的目录,单击 Finish 按钮即可完成 Tomcat 的配置。

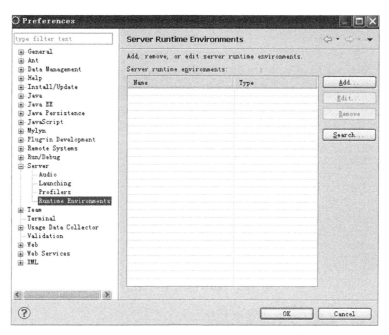

图 1.9　Tomcat 配置界面

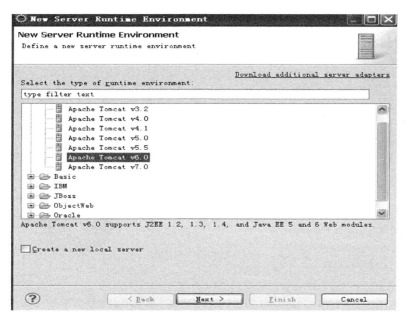

图 1.10　Tomcat 版本选择

4. 任务小结或知识扩展

(1) 软件版本

由于 Java 版本的不断更新，读者下载的 JDK、Tomcat 以及 Eclipse 版本可能和本书使用的不同。但高版本兼容低版本，所以读者可以放心下载和使用最新的版本，这些版本的安装和配置基本一致。

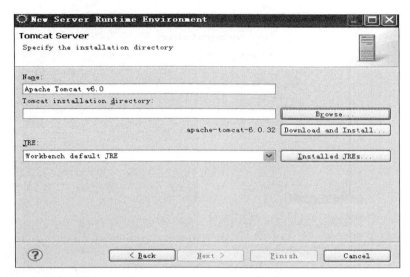

图 1.11　Tomcat 安装目录选择

（2）Windows 7 系统下如何配置系统环境变量

在配置 JDK 的系统环境变量时，对于 Windows Vista 或 Windows 7 系统，只需右击桌面上名为"计算机"的图标，在右键快捷菜单中选择"系统"选项，在弹出的窗口中选择"高级系统设置"选项，同样会弹出"系统属性"对话框，选择名为"高级"的选项卡，单击"环境变量"按钮，最后添加系统环境变量，如图 1.12 所示。

图 1.12　Windows Vista 或 Windows 7 系统下设置环境变量

（3）修改 Tomcat 的默认端口

8080 是 Tomcat 服务器默认占用的端口。但可以通过修改 Tomcat 的配置文件进行端口号修改。用记事本打开 conf 文件夹下的 server.xml 文件，找到以下代码。

```
<Connector port="8080" protocol="HTTP/1.1"
          connectionTimeout="20000"
          redirectPort="8443" />
```

将其中的 port="8080"更改为新的端口号,保存 server.xml 文件后重新启动 Tomcat 服务器即可,比如将 8080 修改为 9090 等。如果修改为 9090,那么在 IE 浏览器地址栏中要输入 http://localhost:9090 才能打开 Tomcat 的测试页面。

需要说明的是,一般情况下,不要修改 Tomcat 默认的端口号,除非 8080 已经被占用。在修改端口时,应避免与公用端口冲突,一旦冲突会影响别的程序正常使用。

1.1.4 实践环节

修改 Tomcat 的端口号并测试。使用记事本打开 Tomcat 目录下的 conf 文件夹中的 server.xml 文件,找到以下代码。

```
<Connector port="8080" protocol="HTTP/1.1"
           connectionTimeout="20000"
           redirectPort="8443" />
```

将其中的 port="8080"更改为新的端口号 9999,保存 server.xml 文件后重新启动 Tomcat 服务器。然后在 IE 浏览器地址栏中输入 http://localhost:9999 打开 Tomcat 的测试页面。

1.2 使用 Eclipse 开发 Web 应用

1.2.1 核心知识

1. JSP 文件

一个 JSP 文件中可以有普通的 HTML 标记和 JSP 规定的 JSP 标记,以及 Java 程序。JSP 文件的扩展名是.jsp,文件的名字必须符合标识符规定,即名字可以由字母、下画线、美元符号和数字组成。

2. JSP 的运行原理

当 Web 服务器上的一个 JSP 页面第一次被客户端请求执行时,Web 服务器上的 JSP 引擎首先将 JSP 文件转译成一个 Java 文件,并将 Java 文件编译成字节码文件,然后执行字节码文件响应客户端的请求。当这个 JSP 页面再次被请求时,JSP 引擎将直接执行字节码文件响应客户端的请求,这也是 JSP 比 ASP 速度快的原因之一。

JSP 引擎以如下方式处理 JSP 页面。
- 将 JSP 页面中的静态元素(HTML 标记)直接交给客户的浏览器执行显示。
- 对 JSP 页面中的动态元素(Java 程序和 JSP 标记)进行必要的处理,将需要显示的结果发送给客户的浏览器。

1.2.2 能力目标

(1) 使用 Eclipse 创建 Web 项目。
(2) 在项目中创建 JSP 文件。
(3) 发布项目到 Tomcat 服务器并运行。

1.2.3 任务驱动

任务的主要内容如下。

1. 创建项目

(1) 启动 Eclipse，进入 Eclipse 的开发界面。

(2) 选择主菜单中的 File→New→Project 菜单项，打开 New Project 对话框，在该对话框中选择 Web 节点下的 Dynamic Web Project 子节点，如图 1.13 所示。

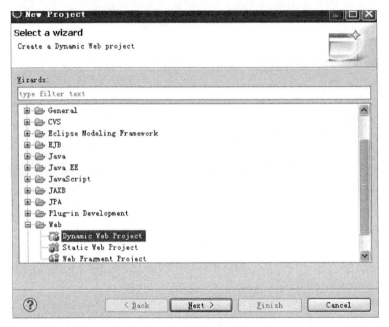

图 1.13　New Project 对话框

(3) 单击 Next 按钮，打开 New Dynamic Web Project 对话框，在该对话框的 Project name 文本框中输入项目名称，这里为 firstTest。选择 Target runtime 区域中的服务器，如图 1.14 所示。

(4) 单击 Finish 按钮，完成项目 firstTest 的创建。此时在 Eclipse 平台左侧的 Project Explorer 中，将显示项目 firstTest，依次展开各节点，可显示如图 1.15 所示的目录结构。

2. 创建 JSP 文件

firstTest 项目创建完成后，可以根据实际需要创建类文件、JSP 文件或者其他文件。这些文件的创建会在需要时介绍，下面将创建一个名字为 myFirst.jsp 的 JSP 文件。

(1) 选中 firstTest 项目的 WebContent 节点并右击，在打开的快捷菜单中，选择 New→JSP File 命令，打开 New JSP File 对话框，在该对话框的 File name 文本框中输入文件名 myFirst.jsp，其他采用默认设置，单击 Finish 按钮完成 JSP 文件的创建。

(2) JSP 创建完成后，在 firstTest 项目的 WebContent 节点下，自动添加一个名称为 myFirst.jsp 的 JSP 文件，同时，Eclipse 会自动将 JSP 文件在右侧的编辑框中打开。

(3) 将 myFirst.jsp 文件中的默认代码修改为以下代码。

```
<%@page language="java" contentType="text/html; charset=GBK" pageEncoding="GBK"%>
```

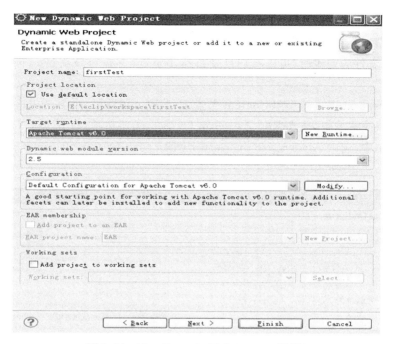

图 1.14 New Dynamic Web Project 对话框

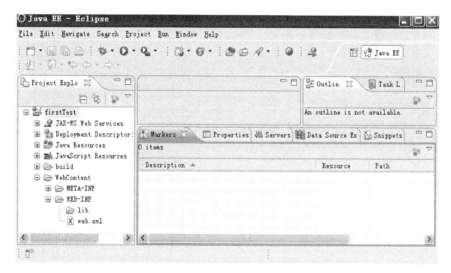

图 1.15 项目 firstTest 的目录结构

```
<html>
<head>
<title>myFirst.jsp</title>
</head>
<body>
<center>真高兴,忙乎半天了,终于要看到人生中第一个 JSP 页面了。</center>
</body>
</html>
```

(4)将编辑好的JSP页面保存(Ctrl+S),至此完成了一个简单的JSP程序的创建。

3. 发布项目到Tomcat并运行

完成JSP文件的创建后,可以将项目发布到Tomcat并运行该项目。下面介绍具体的方法。

(1)在firstTest项目的WebContent节点下,找到myFirst.jsp并选中该JSP文件后右击,在打开的快捷菜单中,选择Run As→Run On Server菜单项,打开Run On Server对话框,在该对话框中,选中Always use this server when running this project复选框,其他采用默认设置,如图1.16所示。

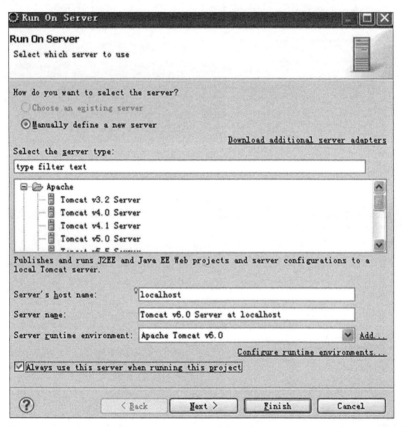

图1.16　Run On Server对话框

(2)单击Finish按钮,即可通过Tomcat运行该项目,运行后的效果如图1.17所示。如果想在IE浏览器中运行该项目,可以将图1.17中的URL地址复制到IE浏览器的地址栏中,并按Enter键运行即可。

4. 任务小结或知识扩展

在创建JSP文件时,Eclipse默认创建的JSP文件的编码格式为ISO-8859-1,为了让页面支持中文,还需要将编码格式修改为GBK或GB2312。

在一个项目的WebContent节点下可以创建多个JSP文件,另外JSP文件中使用到的图片文件、CSS文件(层叠样式表)以及JavaScript文件都放在WebContent节点下。有关

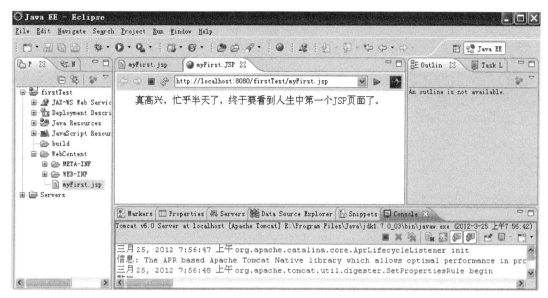

图 1.17　运行 firstTest 项目

CSS 文件和 JavaScript 文件在 JSP 中的应用，读者可以查阅相关书籍。

一个项目中使用的.java 文件都放在 Java Resources 节点下的 src 里面。有关.java 文件在 JSP 中的应用，本书将在第 4 章介绍。

1.2.4　实践环节

（1）参照本节的任务内容 1，创建一个名称为 sencondTest 的项目。

（2）参照本节的任务内容 2，在 sencondTest 项目中创建一个名称为 yourFirst.jsp 的文件，在 JSP 页面中显示"不错！不错！自己能创建 JSP 文件了，并且可以发布运行了。"。

（3）参照本节的任务内容 3，发布并运行 sencondTest 项目。

1.3　小　　结

- JSP 技术不仅是开发 Java Web 应用的先进技术，而且是进一步学习相关技术（如 Struts 框架）的基础。
- JSP 引擎是支持 JSP 程序的 Web 容器，负责运行 JSP 程序，并将有关结果发送给客户端。目前流行的 JSP 引擎有 Tomcat、Resin、JRun、WebSphere、WebLogic 等，本书使用的是 Tomcat 服务器。
- 安装 Tomcat 服务器，首先要安装 JDK，并需要设置 Java_Home 环境变量。
- 当服务器上的一个 JSP 页面第一次被客户端请求执行时，服务器上的 JSP 引擎首先将 JSP 文件转译成一个 Java 文件，并将 Java 文件编译成字节码文件，然后执行字节码文件响应客户端的请求。

习 题 1

1. 安装 Tomcat 服务器所在的计算机需要事先安装 JDK 吗?
2. Tomcat 服务器的默认端口号是什么？如果想修改该端口号,应该修改哪个文件?
3. First.jsp 和 first.jsp 是否是相同的 JSP 文件名字?
4. JSP 引擎是如何处理 JSP 页面中的 HTML 标记的?

第 2 章 JSP 语法

本章主要内容

- JSP 页面的基本构成
- JSP 脚本元素
- JSP 指令标记
- JSP 动作标记

一个 JSP 页面通常是由普通的 HTML 标记、JSP 注释、Java 脚本元素(包括声明、Java 程序片和 Java 表达式)以及 JSP 标记(包括指令标记、动作标记和自定义标记)4 种基本元素组成。这 4 种基本元素在 JSP 页面中是如何被使用的,将是本章介绍的重点。

在本章中,将新建一个 Web 工程 firstTest,本章例子中涉及的 JSP 页面保存在 firstTest 的 WebContent 目录中。

2.1 JSP 页面的基本构成

2.1.1 核心知识

在传统的 HTML 静态页面文件中加入和 Java 相关的动态元素,就构成了一个 JSP 页面。一个 JSP 页面通常由如下 4 种基本元素组成。

(1) 普通的 HTML 标记。
(2) JSP 注释。
(3) Java 脚本元素,包括声明、Java 程序片和 Java 表达式。
(4) JSP 标记,包括指令标记、动作标记和自定义标记。

2.1.2 能力目标

能够识别 JSP 页面的基本元素。

2.1.3 任务驱动

1. 任务的主要内容

根据 example2_1.jsp 代码中的注释,识别 JSP 页面的基本元素。

2. 任务的代码模板

仔细阅读下列 example2_1.jsp 代码,特别注意其中的注释内容。

example2_1.jsp

```jsp
<%@ page language="java" contentType="text/html; charset=GBK" pageEncoding="GBK"%><!--JSP 指令标记 -->
<jsp:include page="a.jsp"/>         <!--JSP 动作标记 -->
<%!
int i=0;                            //变量的声明
int add(int x,int y)                //方法的定义
{
    return x+y;
}
%>
<html>                              <!--html 标记 -->
<head>
<title>example2_1.jsp</title>
</head>
<body>
<%
    i++;                            //Java 程序片
    int result=add(1,2);
%>
i 的值为<%=i%>                       <%--Java 表达式 --%>
<br>
1+2 的和为<%=result%>
</body>
</html>
```

a.jsp

```jsp
<%@page language="java" contentType="text/html; charset=GBK" pageEncoding="GBK"%>
<html>
<head>
<title>a.jsp</title>
</head>
<body>
    被 example2_1.jsp 动态引用
</body>
</html>
```

3. 任务小结或知识扩展

在 example2_1.jsp 代码中,可以看到许多 JSP 注释。注释能够增强 JSP 文件的可读性,便于 Web 项目的更新和维护。JSP 页面中常见的注释有以下两种。

(1) HTML 注释

格式:

```
<!--HTML 注释-->
```

在标记符"<!--"和"-->"之间加入注释内容,就构成了 HTML 注释。

JSP 引擎将 HTML 注释交给客户端,客户通过浏览器查看 JSP 的源文件时,能够看到 HTML 注释。

(2) JSP 注释

格式:

```
<%--JSP 注释--%>
```

在标记符"<%--"和"--%>"之间加入注释内容,就构成了 JSP 注释。

JSP 引擎在编译 JSP 页面时将忽略 JSP 注释,而且客户通过浏览器查看 JSP 的源文件时,无法看到 JSP 注释。

2.1.4 实践环节

识别出如下 JSP 页面的基本元素。

```
<%@page language="java" contentType="text/html; charset=GBK" pageEncoding="GBK"%>
<!--学习 JSP 页面的基本构成 -->
<%!
    String content="JSP 页面基本构成:";
%>
<html>
<head>
<title>shijian2_1.jsp</title>
</head>
<body>
<%
    content=content+"HTML 标记、JSP 注释、JSP 标记以及 Java 脚本元素";
%>
<%=content%>
</body>
</html>
```

2.2 Java 程序片

2.2.1 核心知识

在标记符"<%"和"%>"之间插入的 Java 代码被称做 JSP 页面的 Java 程序片。Java 程序片格式如下:

```
<%Java 代码 %>
```

一个 JSP 页面可以有许多程序片,这些程序片将被 JSP 引擎(本书中指 Tomcat 服务器)按顺序执行。在一个程序片中声明的变量称为 JSP 页面的局部变量,它们在 JSP 页面后继的所有程序片部分以及表达式部分内都有效。

当多个客户请求一个 JSP 页面时,JSP 引擎为每个客户启动一个线程,不同的线程会分别执行该 JSP 页面中的 Java 程序片,程序片中的局部变量会在不同的线程中被分配不同的内存空间。因此,一个客户对 JSP 页面局部变量操作的结果,不会影响到其他客户。Java 程序片执行原理如图 2.1 所示。

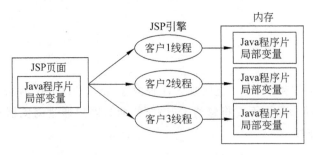

图 2.1　Java 程序片执行

2.2.2　能力目标

理解 Java 程序片的执行原理,掌握在 JSP 页面中如何使用 Java 程序片。

2.2.3　任务驱动

1. 任务的主要内容

编写一个 JSP 页面 example2_2.jsp,该页面在程序片内声明了一个整型的局部变量 x,其初始值为 0。

2. 任务的代码模板

将下列 example2_2.jsp 中的【代码】替换为真正的 JSP 的代码。

example2_2.jsp

```
<%@page language="java" contentType="text/html; charset=GBK" pageEncoding=
"GBK"%>
<html>
<head>
    <title>example2_2.jsp</title>
</head>
    <body>
        【代码 1】                      //Java 程序片开始
        【代码 2】                      //声明 int 型局部变量 x,初始值为 0
        x++;
        out.print("x=" +x);           //在浏览器中输出 x 的值
        【代码 3】                      <!--Java 程序片结束 -->
    </body>
</html>
```

3. 任务小结或知识扩展

如果有 5 个客户请求 example2_2.jsp 页面,JSP 引擎会启动 5 个线程,页面中的 Java 程序片在每个线程中均会被执行一次,共计执行 5 次;在内存中,局部变量 x 对应 5 处不同的存储空间,初始值都为 0,且都只执行了一次自加运算。所以,5 个不同的客户看到的页面效果是相同的,如图 2.2 所示。

图 2.2　example2_2.jsp 页面的执行结果

有时可以根据需要将一个 Java 程序片分割成几个更小的程序片,以便在这些小的程序片之间再插入 JSP 页面的一些其他标记元素。例如,在浏览器中输出大小为 $15×10$ 表格的代码如下:

```
<%@ page language="java"   contentType="text/html; charset=GBK"   pageEncoding=
"GBK"%>
<html>
<head>
    <title>程序片分割</title>
</head>
    <body>
        <table border="1">
<%
    for (int i=1; i <=10; i++)
    {
%>
        <tr>
            <%
                for (int j=1; j <=15; j++)
                {
                    int temp=i * j;
            %>
            <td>
            <%
                out.print(i * j);
            %>
            </td>
            <%
                }
            %>
        </tr>
<%
    }
%>
        </table>
    </body>
</html>
```

4. 代码模板的参考答案

【代码 1】: <%
【代码 2】: int x=0;

【代码 3】：%>

2.2.4 实践环节

编写一个 JSP 页面，在 JSP 页面中使用 Java 程序片输出 26 个小写的英文字母表。页面运行效果如图 2.3 所示。

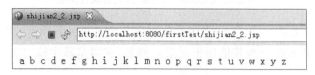

图 2.3 小写英文字母表

2.3 成员变量和方法的定义

在"<％!"和"％>"标记之间可以声明 JSP 的成员变量，定义 JSP 页面的方法。

2.3.1 核心知识

成员变量声明或方法定义的格式如下：

```
<%! 变量声明或方法定义 %>
```

在标记符"<％!"和"％>"之间声明的变量被称做 JSP 页面的成员变量，它们可以是 Java 语言允许的任何数据类型（包括基本数据类型和引用数据类型），例如：

```
<%!
    int n=10;
    Date date;
%>
```

成员变量在整个 JSP 页面内都有效（与书写位置无关），因为 JSP 引擎将 JSP 页面转译成 Java 文件时，将这些变量作为类的成员变量，这些变量的内存空间直到服务器关闭才释放。因此，多个用户共享 JSP 页面的成员变量。也就是说，任何用户对 JSP 页面成员变量操作的结果，都会影响到其他用户。

在标记符"<％!"和"％>"之间定义的方法被称做 JSP 页面的成员方法，该方法在整个 JSP 页面内有效，但是该方法内声明的变量只在该方法内有效，例如：

```
<%!
    int  add(int x,int y) {
      return x+y;
    }
%>
```

2.3.2 能力目标

理解 JSP 成员变量和方法的执行原理，学会使用 JSP 成员变量和方法。

2.3.3 任务驱动

1. 任务的主要内容

编写一个 JSP 页面 example2_3.jsp，页面中声明一个成员变量 n（初始值为 0），并定义一个方法 add（求两个整数的和）。另外，页面中还有一段 Java 程序片，在程序片声明一个局部变量 m，并且对成员变量 n 和局部变量 m 分别进行自加。

2. 任务的代码模板

将下列 example2_3.jsp 中的【代码】替换为真正的 JSP 的代码。

example2_3.jsp

```jsp
<%@page language="java" contentType="text/html; charset=GBK" pageEncoding="GBK"%>
<html>
<head>
    <title>example2_3.jsp</title>
</head>
<%!
    【代码 1】                          //声明成员变量 n,初始值为 0
    int add(int x,int y){              //定义方法
        return x+y;
    }
%>
    <body>
        <%
            【代码 2】                  //声明局部变量 m,初始值为 0
            n++;
            m++;
            int result=add(1,2);
            out.print("成员变量 n 的值为:"+n+"<br>");
            out.print("局部变量 m 的值为:"+m+"<br>");
            out.print("1+2="+result+"<br>"+"<br>");
            out.print("第"+n+"个客户");
        %>
    </body>
</html>
```

3. 任务小结或知识扩展

在 example2_3.jsp 中，变量 n 在标记符"<%!"和"%>"之间声明，因此是成员变量，被所有客户共享；变量 m 在标记符"<%"和"%>"之间声明，因此是局部变量，被每个客户独享。如果有 3 个客户请求这个 JSP 页面，则看到的效果如图 2.4 所示。

从任务中可以看出 Java 程序片具有如下特点。

（1）调用 JSP 页面定义的方法。
（2）操作 JSP 页面声明的成员变量。
（3）声明局部变量。
（4）操作局部变量。

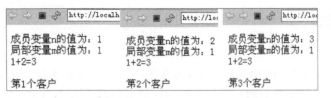

图 2.4　3 个客户请求 example2_3.jsp 页面的效果

4. 代码模板的参考答案

【代码 1】：int n=0;
【代码 2】：int m=0;

2.3.4　实践环节

利用成员变量被所有客户共享这一性质，实现一个简单的计数器，页面效果如图 2.5 所示。

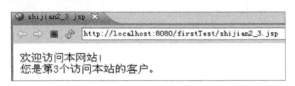

图 2.5　简单的计数器

2.4　Java 表达式

2.4.1　核心知识

在标记符"<%="和"%>"之间可以插入一个表达式，这个表达式必须能求值。表达式的值由 Web 服务器负责计算，并将计算结果用字符串形式发送到客户端，作为 HTML 页面的内容显示。

2.4.2　能力目标

能够灵活使用 Java 表达式计算数据并显示数据信息。

2.4.3　任务驱动

1. 任务的主要内容

在 example2_4.jsp 页面中使用 Java 表达式计算数据并显示数据信息，页面效果如图 2.6 所示。

图 2.6　example2_4.jsp 页面的执行结果

2. 任务的代码模板

将下列 example2_4.jsp 中的【代码】替换为真正的 JSP 的代码。

example2_4.jsp

```jsp
<%@page language="java" contentType="text/html; charset=GBK" pageEncoding="GBK"%>
<html>
<head>
    <title>example2_4.jsp</title>
</head>
<%!
    int add(int x,int y){
        return x+y;
    }
%>
<body>
    1+2 等于【代码 1】 <br>    <!--使用Java表达式调用add方法计算1和2的值-->
    若半径为5,则圆的面积是:【代码 2】 <br> <!--使用Java表达式计算出圆面积-->
    2大于5是否成立:【代码 3】 <br>    <!--使用Java表达式判断1是否大于2-->
</body>
</html>
```

3. 任务小结或知识扩展

Java 表达式中可以有算术表达式、逻辑表达式或条件表达式等。但使用 Java 表达式时,应该注意以下两点。

(1) 不可在"<%="和"%>"之间插入语句,即:输入的内容末尾不能以分号结束。

(2) "<%="是一个完整的符号,"<%"和"="之间不能有空格。

4. 代码模板的参考答案

【代码 1】:`<%=add(1,2) %>`
【代码 2】:`<%=3.14*5*5 %>`
【代码 3】:`<%=1>2 %>`

2.4.4 实践环节

使用 Java 表达式显示出系统的当前时间。页面效果如图 2.7 所示。

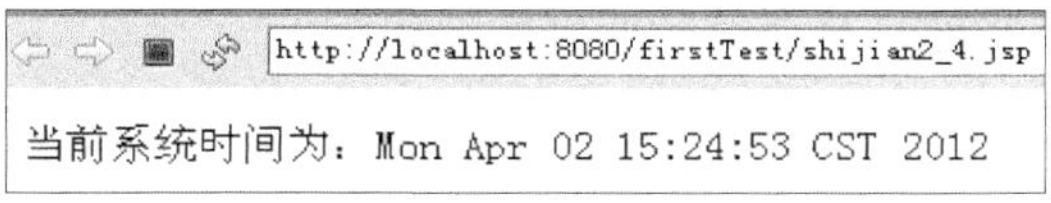

图 2.7 显示系统的当前时间

2.5 page 指令标记

page 指令标记用来定义整个 JSP 页面的一些属性和这些属性的值。可以用一个 page 指令指定多个属性的值,也可以使用多个 page 指令分别为每个属性指定值。page 指令的格式如下：

```
<%@page 属性1="属性1的值"  属性2="属性2的值" …%>
```

或者

```
<%@page 属性1="属性1的值" %>
<%@page 属性2="属性2的值" %>
<%@page 属性3="属性3的值" %>
   ⋮
<%@page 属性n="属性n的值" %>
```

page 指令的主要属性有 contentType、import、language 和 pageEncoding 等。

2.5.1 核心知识

page 指令标记通常定义的属性有以下 4 种。

1. 属性 contentType

JSP 页面使用 page 指令标记只能为 contentType 属性指定一个值，用来确定响应的 MIME 类型（MIME 类型就是设定某种文件用对应的一种应用程序来打开的方式类型）。当用户请求一个 JSP 页面时，服务器会告诉客户的浏览器使用 contentType 属性指定的 MIME 类型来解释执行所接收到的服务器为之响应信息。如果希望客户的浏览器使用 Word 应用程序打开用户请求的页面，就可以把 contentType 属性的值设置为

```
<%@page contentType="application/ msword;charset=GBK"%>
```

2. 属性 import

JSP 页面使用 page 指令标记可为 import 属性指定多个值，import 属性的作用是为 JSP 页面引入包中的类，以便在 JSP 页面的程序片、变量及方法声明或表达式中使用包中的类。

3. 属性 language

language 属性用来指定 JSP 页面使用的脚本语言，目前该属性的值只能取 java。

4. 属性 pageEncoding

contentType 中的 charset 是指服务器发送给客户浏览器时所见到的网页内容的编码；pageEncoding 是指 JSP 文件自身存储时所用的编码。

2.5.2 能力目标

读懂 page 指令标记为 JSP 页面指定的一些属性值。

2.5.3 任务驱动

1. 任务的主要内容

编写一个 JSP 页面 example2_5.jsp，当用户请求该页面时，客户浏览器启动本地的 PowerPoint 应用程序打开该页面。

2. 任务的代码模板

将下列 example2_5.jsp 中的【代码】替换为真正的 JSP 的代码。

example2_5.jsp

```
<%@page
【代码1】="java" <!--代码1设置 language 属性-->
【代码2】="application/vnd.ms-powerpoint; charset=GBK" pageEncoding="GBK"%>
<!--代码2设置 contentType 属性-->
<%@page【代码3】="java.util.*"%>  <!--代码3设置 import 属性-->
<%@page【代码4】="java.io.*"%>  <!--代码4设置 import 属性-->
<html>
<head>
    <title>example2_5.jsp</title>
</head>
<body>
    在学习 page 指令标记时,请牢牢记住只能为 JSP 页面设置一个 contentType 属性值,可为
    import 属性设置多个值.
</body>
</html>
```

3. 任务小结或知识扩展

使用 page 指令为 contentType 属性指定的 MIME 类型,常见的有：text/html(HTML 解析器,所谓的网页形式)、text/plain(普通文本)、application/pdf(PDF 文档)、application/msword(Word 应用程序)、image/jpeg(JPEG 图形)、image/png(PNG 图像)、image/gif(GIF 图形)以及 application/vnd.ms-powerpoint(PowerPoint 应用程序)。

在 JSP 标准的语法中,如果 pageEncoding 属性存在,那么 JSP 页面的字符编码方式就由 pageEncoding 决定;否则就由 contentType 属性中的 charset 决定,如果 charset 也不存在,JSP 页面的字符编码方式就采用默认的 ISO-8859-1。

4. 代码模板的参考答案

【代码1】:language
【代码2】:contentType
【代码3】:import
【代码4】:import

2.5.4 实践环节

把任务中 example2_5.jsp 页面的 contentType 属性值指定为 application/msword,运行修改后的页面,并仔细观察运行结果。

2.6 include 指令标记

2.6.1 核心知识

一个网站中的多个 JSP 页面有时需要显示同样的信息,比如该网站的 Logo 或导航条等,为了便于网站项目的维护,通常在这些 JSP 页面的适当位置嵌入一个相同的文件。

include 指令标记的作用就是把 HTML 网页文件或其他文本文件静态嵌入当前的 JSP 网页中,该指令的语法格式如下:

```
<%@include file="文件的URL"%>
```

所谓静态嵌入,就是"先包含后处理",在编译阶段完成对文件的嵌入。即:先将当前 JSP 页面与要嵌入的文件合并成一个新的 JSP 页面,然后再由 JSP 引擎将新页面转化为 Java 文件处理并运行。

2.6.2　能力目标

理解静态嵌入的概念,并能够使用 include 指令标记在 JSP 网页中静态嵌入文件。

2.6.3　任务驱动

1. 任务的主要内容

编写两个 JSP 页面 example2_6.jsp 和 example2_6_1.jsp,在 example2_6.jsp 页面中使用 include 指令标记静态嵌入 example2_6_1.jsp 页面,访问 example2_6.jsp 页面,运行效果如图 2.8 所示。

图 2.8　include 指令标记的使用

2. 任务的代码模板

将下列 example2_6.jsp 中的【代码】替换为真正的 JSP 的代码。

example2_6.jsp

```
<%@page language="java" contentType="text/html; charset=GBK" pageEncoding="GBK"%>
<html>
<head>
    <title>example2_6.jsp</title>
</head>
<body>
    静态嵌入 example2_6_1.jsp 之前
    <br>
    【代码】  <!--使用 include 静态嵌入 example2_6_1.jsp-->
    <br>
    静态嵌入 example2_6_1.jsp 之后
</body>
</html>
```

example2_6_1.jsp

```
<%@page language="java" contentType="text/html; charset=GBK" pageEncoding="GBK"%>
```

```
<html>
<head>
<title>example2_6_1.jsp</title>
</head>
<body>
    <font color="red" size=4>example2_6_1.jsp 文件的内容</font>
</body>
</html>
```

3. 任务小结或知识扩展

在该任务中，example2_6.jsp 页面静态嵌入 example2_6_1.jsp 页面，此时需要先将 example2_6_1.jsp 中的所有代码全部嵌入 example2_6.jsp 的指定位置，形成一个新的 JSP 文件，然后再将新文件提交给 JSP 引擎处理，如图 2.9 所示。

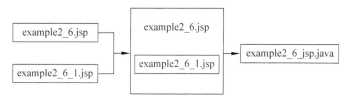

图 2.9 静态嵌入的原理

在使用 include 指令标记时，需要注意嵌入文件后必须保证新合成的 JSP 页面符合 JSP 的语法规则，比如该任务中的 example2_6.jsp 和 example2_6_1.jsp 两个页面的 page 指令就不能指定不同的 contentType 值，否则在合并后的 JSP 页面就两次使用 page 指令为 contentType 属性设置了不同的属性值，导致语法错误。

4. 代码模板的参考答案

【代码】:`<%@include file="example2_6_1.jsp" %>`

2.6.4 实践环节

将任务中 example2_6.jsp 页面的 contentType 属性值修改为 "application/msword; charset=GBK"，并运行修改后的页面。

2.7 include 动作标记

2.7.1 核心知识

动作标记 include 的作用是把 JSP 文件、HTML 网页文件或其他文本文件动态嵌入当前的 JSP 网页中，该指令的语法有以下两种格式。

`<jsp:include  page="文件的 URL"/>`

或

`<jsp:include  page="文件的 URL">`

子标记

```
<jsp:include/>
```

关于子标记的问题本书将在2.9节详细介绍，当动作标记 include 不需要子标记时，必须使用上述第一种形式。

所谓动态嵌入，就是"先处理后包含"，在运行阶段完成对文件的嵌入。即：在把 JSP 页面转译成 Java 文件时，并不合并两个页面；而是在 Java 文件的字解码文件被加载并执行时，才去处理 include 动作标记中引入的文件。与静态嵌入方式相比，动态嵌入的执行速度稍慢，但是灵活性较高。

2.7.2 能力目标

理解动态嵌入的概念，并能够使用 include 动作标记在 JSP 网页中动态嵌入文件。

2.7.3 任务驱动

1. 任务的主要内容

编写两个 JSP 页面 example2_7.jsp 和 example2_7_1.jsp，在 example2_7.jsp 页面中使用 include 动作标记动态嵌入 example2_7_1.jsp 页面。运行 example2_7.jsp 页面。

2. 任务的代码模板

将下列 example2_7.jsp 中的【代码】替换为真正的 JSP 的代码。

example2_7.jsp

```
<%@page language="java" contentType="text/html; charset=GBK" pageEncoding="GBK"%>
<html>
<head>
    <title>example2_7.jsp</title>
</head>
<body>
    动态嵌入 example2_7_1.jsp 之前
    <br>
    【代码】  <!--使用 include 动作标记动态嵌入 example2_7_1.jsp-->
    <br>
    动态嵌入 example2_7_1.jsp 之后
</body>
</html>
```

example2_7_1.jsp

```
<%@page language="java" contentType="text/html; charset=GBK" pageEncoding="GBK"%>
<html>
<head>
    <title>example2_7_1.jsp</title>
</head>
<body>
    <font color="red" size=4>example2_7_1.jsp 文件的内容</font>
```

```
</body>
</html>
```

3. 任务小结或知识扩展

在该任务中,文件 example2_7.jsp 通过动作标记 include 动态嵌入了文件 example2_7_1.jsp,此时 JSP 引擎不会将两个文件合并成一个 JSP 页面,而是分别将 example2_7.jsp 文件和 example2_7_1.jsp 文件转化成对应的 Java 文件和字节码文件。当 JSP 解释器解释执行 example2_7.jsp 页面时,会遇到动作指令＜jsp:include page="example2_7_1.jsp"/＞对应的代码,此时才会执行 example2_7_1.jsp 页面对应的字节码文件,然后将执行的结果发送到客户端,并由客户端负责显示这些结果,所以 example2_7.jsp 和 example2_7_1.jsp 页面中 page 指令的 contentType 属性值可以不同。

4. 代码模板的参考答案

【代码】:`<jsp:include page="example2_7_1.jsp" />`

2.7.4 实践环节

将任务中 example2_7.jsp 页面的 contentType 属性值修改为"application/msword;charset＝GBK",并运行修改后的页面。

2.8 forward 动作标记

2.8.1 核心知识

动作标记 forward 的作用是:从该标记出现处停止当前 JSP 页面的继续执行,从而转向执行 forward 动作标记中 page 属性值指定的 JSP 页面。该标记有两种格式。

```
<jsp: forward  page="文件的 URL"/>
```

或

```
<jsp: forward  page="文件的 URL">
    子标记
<jsp: forward />
```

当动作标记 forward 不需要子标记时,必须使用上述第一种形式。

2.8.2 能力目标

能够使用 forward 动作标记在 JSP 网页中实现页面的跳转。

2.8.3 任务驱动

1. 任务的主要内容

编写 example2_8.jsp、evenNumbers.jsp 和 oddNumber.jsp 3 个 JSP 页面。在 example2_8.jsp 页面中使用 forward 动作标记转向 evenNumbers.jsp 或 oddNumber.jsp 页面,在

example2_8.jsp 页面中随机获取 0~10 之间的整数,该数为偶数就转向页面 evenNumbers.jsp,否则转向页面 oddNumber.jsp。首先访问 example2_8.jsp 页面。

2. 任务的代码模板

将下列 example2_8.jsp 中的【代码】替换为真正的 JSP 的代码。

example2_8.jsp

```jsp
<%@page language="java" contentType="text/html; charset=GBK" pageEncoding="GBK"%>
<html>
<head>
<title>example2_8.jsp</title>
</head>
<body>
<%
    long i=Math.round(Math.random() * 10);
    if(i%2==0){
        System.out.println("获得的整数是偶数,即将跳转到偶数页面 evenNumbers.jsp。");
%>
    【代码 1】<!--使用 forward 动作标记转向 evenNumbers.jsp 页面-->
<%
    System.out.println("我是偶数尝试一下能看到我吗?");
    }
    else
    {
        System.out.println("获得的整数是奇数,即将跳转到奇数页面 oddNumber.jsp。");
%>
    【代码 2】<!--使用 forward 动作标记转向 oddNumber.jsp 页面-->
<%
        System.out.println("我是奇数尝试一下能看到我吗?");
    }
%>
</body>
</html>
```

evenNumbers.jsp

```jsp
<%@page language="java" contentType="text/html; charset=GBK" pageEncoding="GBK"%>
<html>
<head>
<title>evenNumbers.jsp</title>
</head>
<body>
    我是偶数页。
</body>
</html>
```

oddNumber.jsp

```jsp
<%@page language="java" contentType="text/html; charset=GBK" pageEncoding=
```

```
"GBK"%>
<html>
<head>
<title>oddNumber.jsp</title>
</head>
<body>
    我是奇数页。
</body>
</html>
```

3. 任务小结或知识扩展

在该任务中，当用户请求查看页面 example2_8.jsp 时，如果获取的整数是偶数，那么只会在控制台上看到"获得的整数是偶数，即将跳转到偶数页面 evenNumbers.jsp。"这句话，当 JSP 引擎执行到＜jsp:forward page＝"evenNumbers.jsp" /＞语句时，会停止当前页面的执行，然后自动跳转到 evenNumbers.jsp 页面，并在客户端的浏览器上显示 evenNumbers.jsp 页面中的内容。如果获取的整数是奇数，那么只会在控制台上看到"获得的整数是奇数，即将跳转到奇数页面 oddNumber.jsp。"这句话，当 JSP 引擎执行到＜jsp:forward page＝"oddNumber.jsp" /＞语句时，会停止当前页面的执行，然后自动跳转到 oddNumber.jsp 页面，并在客户端的浏览器上显示 oddNumber.jsp 页面中的内容。

4. 代码模板的参考答案

【代码 1】:`<jsp:forward page="evenNumbers.jsp"/>`
【代码 2】:`<jsp:forward page="oddNumber.jsp"/>`

2.8.4 实践环节

将 example2_8.jsp 页面中的 forward 动作标记修改为 include 动作标记，并对修改前与修改后的运行结果进行比较。

2.9 param 动作标记

2.9.1 核心知识

动作标记 param 不能独立使用，但可以作为 include、forward 动作标记的子标记来使用，该标记以"名字-值"对的形式为对应页面传递参数。该标记的格式为

```
<jsp:父标记  page="接收参数页面的 URL">
    <jsp:param  name="参数名"  value="参数值"/>
<jsp:父标记/>
```

接收参数的页面可以使用内置对象 request 调用 getParameter("参数名")方法获取动作标记 param 传递过来的参数值，内置对象将在本书第 3 章介绍。

2.9.2 能力目标

能够使用 param 动作标记作为 include、forward 动作标记的子标记为对应页面传递

参数。

2.9.3 任务驱动

1. 任务的主要内容

编写 example2_9.jsp 和 show.jsp 两个 JSP 页面，在 example2_9.jsp 页面中使用 include 动作标记动态包含文件 show.jsp，并向它传递一个名为 userName，值为 kazhafei 的参数；show.jsp 收到参数后，计算参数值的字符个数，并输出参数值和它的字符个数。运行 example2_9.jsp 页面，效果如图 2.10 所示。

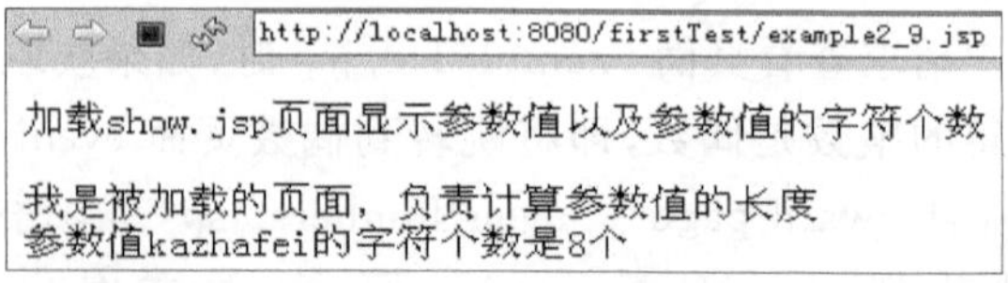

图 2.10 用 param 子标记向加载的文件传值

2. 任务的代码模板

将下列 example2_9.jsp 中的【代码】替换为真正的 JSP 的代码。

example2_9.jsp

```jsp
<%@page language="java" contentType="text/html; charset=GBK" pageEncoding="GBK"%>
<html>
<head>
<title>example2_9.jsp</title>
</head>
<body>
    加载 show.jsp 页面显示参数值以及参数值的字符个数<br>
    <jsp:include page="show.jsp">
        【代码】
    </jsp:include>
<!-- 【代码】使用 param 子标记传递一个名为 userName，值为 kazhafei 的参数 -->
</body>
</html>
```

show.jsp

```jsp
<%@page language="java" contentType="text/html; charset=GBK" pageEncoding="GBK"%>
<html>
<head>
<title>show.jsp</title>
</head>
<body>
    <%
        String name=request.getParameter("userName");
        int n=name.length();
        out.print("我是被加载的页面,负责计算参数值的长度<br>");
```

```
        out.print("参数值"+name+"的字符个数是"+n+"个");
    %>
</body>
</html>
```

3. 任务小结或知识扩展

当使用 include 动作标记时经常会使用 param 子标记，以便向动态包含的 JSP 文件传递必要的参数值，这也体现出 include 动作标记比 include 指令标记灵活的特点。如果向页面传递多个参数，可以多次使用 param 子标记。格式如下：

```
<jsp:父标记  page="接收参数页面的URL">
    <jsp:param  name="参数名1"  value="参数值1"/>
    <jsp:param  name="参数名2"  value="参数值2"/>
    <jsp:param  name="参数名3"  value="参数值3"/>
       ⋮
<jsp:父标记/>
```

4. 代码模板的参考答案

【代码】：`<jsp:param value="kazhafei" name="userName"/>`

2.9.4 实践环节

编写 shijian2_9.jsp 和 computer.jsp 两个页面，在页面 shijian2_9.jsp 中使用 include 动作标记动态包含文件 computer.jsp，并向它传递一个矩形的长和宽；computer.jsp 收到参数后，计算矩形的面积，并显示结果。运行 shijian2_9.jsp 页面，效果如图 2.11 所示。

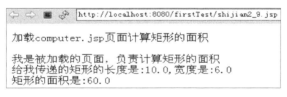

图 2.11 用 param 子标记向加载的文件传递多个值

2.10 小　　结

- 一个 JSP 页面通常由普通的 HTML 标记、JSP 注释、Java 脚本元素（包括声明、Java 程序片和 Java 表达式）以及 JSP 标记（包括指令标记、动作标记和自定义标记）组成。
- 在一个程序片中声明的变量称为 JSP 页面的局部变量，它们在 JSP 页面后继的所有程序片部分以及表达式部分内都有效。一个用户对 JSP 页面局部变量操作的结果，不会影响到其他用户。
- JSP 页面成员变量是被所有用户共享的变量，任何用户对 JSP 页面成员变量操作的结果，都会影响到其他用户。

- page 指令标记用来定义整个 JSP 页面的一些属性和这些属性的值。page 指令只能为 contentType 属性指定一个值,但可为 import 属性指定多个值。
- include 指令标记是先将当前 JSP 页面与要嵌入的文件合并成一个新的 JSP 页面,然后再由 JSP 引擎将新页面转化为 Java 文件处理并运行。而 include 动作标记在把 JSP 页面转译成 Java 文件时,并不合并两个页面;而是在 Java 文件的字解码文件被加载并执行时,才去处理 include 动作标记中引入的文件。

习 题 2

1. JSP 页面中由哪几种主要元素组成?
2. 如果有 3 个用户访问一个 JSP 页面,则该页面中的 Java 程序片将被执行几次?
3. "<%!"和"%>"之间声明的变量与"<%"和"%>"之间声明的变量有何不同?
4. 动作标记 include 和指令标记 include 的区别是什么?
5. 一个 JSP 页面中是否允许使用 page 指令为 contentType 属性设置多个值?是否允许使用 page 指令为 import 属性设置多个值?
6. 编写 3 个 JSP 页面:main.jsp、first.jsp 和 second.jsp,将 3 个 JSP 文件保存在同一个 Web 服务目录中,main.jsp 使用 include 动作标记加载 first.jsp 和 second.jsp 页面。first.jsp 页面可以画出一张表格,second.jsp 页面可以计算出两个正整数的最大公约数。当 first.jsp 被加载时获取 main.jsp 页面 include 动作标记的 param 子标记提供的表格的行数和列数,当 second.jsp 被加载时获取 main.jsp 页面 include 动作标记的 param 子标记提供的两个正整数的值。

第 3 章 JSP 内置对象

本章主要内容

- 请求对象 request
- 响应对象 response
- 会话对象 session
- 全局应用程序对象 application

有些对象在 JSP 页面中不需要声明和实例化,可以直接在 Java 程序片和 Java 表达式部分使用,这些对象就是 JSP 的内置对象。内置对象由 Web 服务器负责实现和管理。

常见的 JSP 内置对象有 request、response、session、application 以及 out。在本章中将主要学习这几种内置对象的使用方法。

在本章中,将新建一个 Web 工程 ch3,本章例子中涉及的 JSP 页面保存在 ch3 的 WebContent 目录中。

3.1 请求对象 request

3.1.1 核心知识

request 内置对象是实现了 javax.servlet.ServletRequest 接口的一个实例。当用户请求一个 JSP 页面时,JSP 页面所在的服务器将用户发出的所有请求信息封装在内置对象 request 中,使用该对象就可以获取用户提交的信息。

request 对象获取客户提交信息的两个常用方法如下。

1. public String getParameter(String name)

该方法以字符串的形式返回客户端传来的某个参数的值,该参数名由 name 指定。例如:

```
<form action="getValue.jsp">
    <input type="text" name="userName"/>
    <input type="submit" value="提交"/>
</form>
```

单击"提交"按钮向 JSP 页面 getValue.jsp 提交信息后就可以获得文本框中输入的信息,代码如下：

```
String name=request.getParameter("userName");
```

2. public String[] getParameterValues(String name)

该方法以字符串数组的形式返回客户端向服务器端传递的指定参数名的所有值。例如：

```
<form action="getValues.jsp">
    选择你去过的城市:<br/>
    <input type="checkbox" name="cities" value="Beijing"/>北京
    <input type="checkbox" name="cities" value="Shanghai"/>上海
    <input type="checkbox" name="cities" value="Hong Kong"/>香港
    <input type="submit" value="提交"/>
</form>
```

如果选择了北京和上海两个城市，提交表单后可以用 getParameterValues 方法，以复选框的 name 属性值"cities"为参数，获取到一个数组，其中元素分别对应北京和上海两个选项的 value 值"Beijing"和"Shanghai"。代码如下：

```
String yourCities[]=request.getParameterValues("cities");
```

3.1.2 能力目标

能够灵活使用 request 内置对象获取客户提交的信息。

3.1.3 任务驱动

1. 任务的主要内容

编写 example3_1.jsp 和 getValue.jsp 两个 JSP 页面，example3_1.jsp 通过表单向 getValue.jsp 提交输入的姓名和选择的城市，getValue.jsp 负责获得表单中提交的信息并显示。页面运行效果如图 3.1 所示。

图 3.1　example3_1.jsp 的效果图

2. 任务的代码模板

将下列 example3_1.jsp 中的【代码】替换为真正的 JSP 的代码。

example3_1.jsp

```
<%@page language="java" contentType="text/html; charset=GBK" pageEncoding="GBK"%>
<html>
<head>
<title>example3_1.jsp</title>
```

```
</head>
<body>
    <form action="getValue.jsp">
        姓名:<input type="text" name="userName"/>
        <br>
        选择您去过的城市:
        <input type="checkbox" name="cities" value="Beijing"/>北京
        <input type="checkbox" name="cities" value="Shanghai"/>上海
        <input type="checkbox" name="cities" value="Hong Kong"/>香港
        <input type="checkbox" name="cities" value="Dalian"/>大连
        <br>
        <input type="submit" value="提交"/>
    </form>
</body>
</html>
```

getValue.jsp

```
<%@page language="java" contentType="text/html;charset=GBK" pageEncoding="GBK"%>
<html>
<head>
<title>getValue.jsp</title>
</head>
<body>
<%
    String name=【代码1】           //获得example3_1.jsp页面中输入的姓名
    String cities[]=【代码2】        //获得example3_1.jsp页面中选择的城市
%>
    您输入的姓名是:<%=name %><br>
    您去过的城市有:
    <%
        for(int i=0;i<cities.length;i++){
            out.print(cities[i]+" ");
        }
    %>
</body>
</html>
```

3. 任务小结或知识扩展

1) NullPointerException 异常

如果不选择 example3_1.jsp 页面的城市,而直接单击"提交"按钮,那么 getValue.jsp 页面就会提示出现 NullPointerException 异常。为了避免在运行时出现 NullPointerException 异常,我们在 getValue.jsp 页面中使用如下代码。

```
if(cities!=null){
    for(int i=0;i<cities.length;i++){
        out.print(cities[i]+" ");
    }
}
```

2）中文乱码问题

如果在 example3_1.jsp 页面的文本框中输入中文姓名，那么 getValue.jsp 页面获得的姓名是乱码（由很多"?"组成）。此时必须对含有汉字字符的信息进行特殊的处理。乱码解决常用的方法有如下两种。

（1）使用 setCharacterEncoding(String code) 设置统一字符编码

request 对象提供了方法 setCharacterEncoding(String code) 设置编码，其中参数 code 以字符串形式传入要设置的编码格式，但这种方法仅对于提交方式是 post 的表单（表单默认的提交方式是 get）有效。例如，我们使用该方法解决 example3_1.jsp 和 getValue.jsp 页面中出现的中文乱码问题，需要完成如下两件事。

首先，将 example3_1.jsp 中的表单提交方式改为"post"，具体代码如下：

```
<form action="getValue.jsp" method="post">
```

其次，在 getValue.jsp 中获取表单信息之前设置统一编码，具体代码如下：

```
request.setCharacterEncoding("GBK");
```

（2）对获取的信息进行重新编码

通过内置对象 request 获取到字符串的值后，对该字符串使用 ISO-8859-1 重新编码，并把编码的结果存放到一个字节数组中，然后再把这个字节数组转化为字符串。例如，我们使用该方法解决 example3_1.jsp 和 getValue.jsp 页面中出现的中文乱码问题，具体代码如下：

```
String name=request.getParameter("userName");
byte b[]=name.getBytes("ISO-8859-1");
name=new String(b);
```

3）字符集

为了更好地理解中文乱码的解决方案，需要了解几种常用的字符集。

（1）ASCII。ASCII(American Standard Code for Information Interchange，美国信息互换标准代码)，是基于常用英文字符的一套编码。

（2）ISO-8859-1。ISO-8859-1 编码通常叫做 Latin-1，除收录 ASCII 字符外，还增加了其他一些语言和地区需要的字符。该编码是 Tomcat 服务器默认采用的字符编码。

（3）GB2312。GB2312 码是中华人民共和国国家标准汉字信息交换用编码，简称国标码，是由国家标准总局发布的关于汉字的编码，通行于中国内地和新加坡。

（4）GBK。GBK 编码规范，除了完全兼容 GB2312 外，还对繁体中文和一些不常用的字符进行了编码。GBK 是现阶段 Windows 和其他一些中文操作系统的默认字符集。

（5）Unicode。Unicode 为统一的字符编码标准集，为地球上几乎所有地区每种语言中的每个字符设定了统一并且唯一的编码，以满足跨语言、跨平台进行文本转换、处理的要求。

（6）UTF-8。UTF-8 是 Unicode 的一种变长字符编码。用在网页上可以同一页面显示中文和其他语言。当处理包含多国文字的信息页面时一般选择用 UTF-8。

4）request 的常用方法

request 的常用方法除了核心知识中的两个外，还有如下几个。

（1）public void setAttribute(String name,Object obj)

该方法可以将某个参数和它的值与当前的 request 对象绑定。name 为参数的名称，obj 为对应的参数值，必须是复合类型的对象。

（2）public Object getAttribute(String name)

该方法返回之前调用 setAttribute 方法时所设置参数 name 对应的属性值，如果对应的属性值不存在，则会返回 null；如果对应的属性值存在，则返回一个 Object 对象，所以需要进行强制类型转换。

（3）public void removeAttribute(String name)

该方法从 request 对象中删除参数 name 所对应的属性。

4. 代码模板的参考答案

【代码 1】:request.getParameter("userName");
【代码 2】:request.getParameterValues("cities");

3.1.4 实践环节

使用两种方法（设置统一编码和重新编码）解决 example3_1.jsp 和 getValue.jsp 页面中出现的中文乱码问题。

3.2 响应对象 response

3.2.1 核心知识

当用户请求服务器的一个页面时，会提交一个 HTTP 请求，服务器收到请求后，返回 HTTP 响应。request 对象对请求信息进行封装，与 request 对象对应的对象是 response 对象。response 对象对用户的请求作出动态响应。动态响应通常有如下 3 个。

1. 动态改变 contentType 属性值

JSP 页面用 page 指令标记设置了页面的 contentType 属性值，response 对象就按照这种属性值的方式对客户作出响应。在 page 指令中只能为 contentType 属性指定一个值。如果想动态改变 contentType 属性值，换一种方式来响应客户，可以让 response 对象调用 setContentType(String s)方法来重新设置 contentType 的属性值。

2. 设置响应表头(HTTP 文件头)

response 对象可以通过方法 setHeader(String name,String value)设置指定名字的 HTTP 文件头的值，以此来操作 HTTP 文件头。response 对象设置的新值将会覆盖原值。如果希望某页面每 3 秒钟刷新一次，那么在该页面中添加如下代码。

response.setHeader("refresh","3");

3. response 重定向

在需要将用户引导至另一个页面时，可以使用 response 对象的 sendRedirect(String url)方法实现用户的重定向。例如，用户输入的表单信息不完整，应该再次被重定向到输入页面。

3.2.2 能力目标

能够灵活使用 response 内置对象动态响应用户的请求。

3.2.3 任务驱动

本节有以下 3 个任务。

1. 任务 1——动态改变 contentType 属性值

（1）任务 1 的主要内容

编写一个 JSP 页面 example3_2_1.jsp，客户通过单击页面上的不同按钮，可以改变页面响应的 MIME 类型。当单击 word 按钮时，JSP 页面动态地改变 contentType 的属性值为 application/msword，客户浏览器启用本地的 Word 软件来显示当前页面内容；当单击 excel 按钮时，JSP 页面动态地改变 contentType 的属性值为 application/vnd.ms-excel，客户浏览器启用本地的 Excel 软件来显示当前页面内容。效果如图 3.2(a)～图 3.2(c)所示。

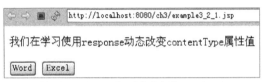

(a) text/html响应方式

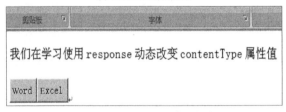

(b) application/msword响应方式

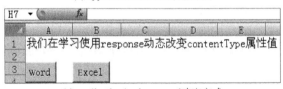

(c) application/vnd.ms-excel响应方式

图 3.2　example3_2_1.jsp 的效果图

（2）任务 1 的代码模板

将下列 example3_2_1.jsp 中的【代码】替换为真正的 JSP 的代码。

example3_2_1.jsp

```
<%@page language="java" contentType="text/html; charset=GBK" pageEncoding="GBK"%>
<html>
<head>
<title>example3_2_1.jsp</title>
```

```
</head>
<body>
  <form action="" method="post">
    <p>我们在学习使用 response 动态改变 contentType 属性值
    <p>
    <input type="submit" value="word" name="submit">
    <input type="submit" value="excel" name="submit">
    <%
        String str=request.getParameter("submit");
        if ("word".equals(str)) {
        【代码 1】
            //response 调用 setContentType 方法设置 MIME 类型为 application/msword
        }else if ("excel".equals(str)) {
        【代码 2】
            //response 调用 setContentType 方法设置 MIME 类型为 application/vnd.ms-excel
        }
    %>
  </form>
</body>
</html>
```

(3) 任务 1 小结或知识扩展

response 对象调用 setContentType(String s)方法来重新设置网页响应的 MIME 类型。常见的 MIME 类型有 text/html、application/msword、application/vnd. ms-excel、image/gif、image/jpeg、application/vnd. ms-powerpoint、application/x-shockwave-flash、application/pdf 等。

(4) 任务 1 代码模板的参考答案

【代码 1】:response.setContentType("application/msword");
【代码 2】:response.setContentType("application/vnd.ms-excel");

2. 任务 2——设置响应表头

(1) 任务 2 的主要内容

编写一个 JSP 页面 example3_2_2.jsp，在该页面中使用 response 对象设置一个响应头"refresh"，其值是"3"。那么用户收到这个头之后，该页面会每 3 秒钟刷新一次。

(2) 任务 2 的代码模板

将下列 example3_2_2.jsp 中的【代码】替换为真正的 JSP 的代码。

example3_2_2.jsp

```
<%@page language="java" contentType="text/html; charset=GBK" pageEncoding="GBK"%>
<%@page import="java.util.*" %>
<html>
<head>
<title>example3_2_2.jsp</title>
</head>
<body>
  <h2>该页面每 3 秒钟刷新 1 次</h2>
```

```
<p>现在的秒钟时间是：
<%
    Date d=new Date();
    out.print(""+d.getSeconds());
    【代码】              //使用response对象设置一个响应头"refresh",其值是"3".
%>
</body>
</html>
```

(3) 任务 2 小结或知识扩展

有时候希望从当前页面几秒钟后自动跳转到另一个页面。比如，打开 a.jsp 页面 3 秒钟后，自动跳转到 b.jsp 页面(a.jsp 与 b.jsp 在同一个 Web 服务目录下)。这该如何实现呢？我们只需为 a.jsp 设置一个响应头即可，也就是在 a.jsp 页面中添加如下代码。

`response.setHeader("refresh","3;url=b.jsp");`

(4) 任务 2 代码模板的参考答案

【代码】：`response.setHeader("refresh","3");`

3. 任务 3——重定向

(1) 任务 3 的主要内容

编写 example3_2_3.jsp 和 enter.jsp 两个 JSP 页面，如果在页面 example3_2_3.jsp 中输入正确的密码"kazhafei"，单击"让我进入安全洞"按钮后提交给页面 enter.jsp；如果输入不正确，重定向到 example3_2_3.jsp 页面。先运行 example3_2_3.jsp 页面，页面效果如图 3.3(a)、图 3.3(b)所示。

(a) example3_2_3.jsp 页面

(b) enter.jsp 页面

图 3.3　example3_2_3.jsp 的效果图

(2) 任务 3 的代码模板

将下列 example3_2_3.jsp 中的【代码】替换为真正的 JSP 的代码。

example3_2_3.jsp

```
<%@page language="java" contentType="text/html; charset=GBK" pageEncoding="GBK"%>
<html>
<head>
<title>example3_2_3.jsp</title>
</head>
<body>
```

```html
<form action="enter.jsp" method="post" name=form>
    <p>
        输入密钥：
    <br>
    <input type="password" name="pwd"/>
    <input type="submit" value="让我进入安全洞">
</form>
</body>
</html>
```

enter.jsp

```jsp
<%@page language="java" contentType="text/html; charset=GBK" pageEncoding="GBK"%>
<html>
<head>
<title>enter.jsp</title>
</head>
<body>
<%
    String str=request.getParameter("pwd");
    if (!"kazhafei".equals(str)) {
        【代码】//重定向到 example3_2_3.jsp 页面重新输入密码
    } else {
        out.print("菲菲,欢迎您!");
    }
%>
</body>
</html>
```

(3) 任务 3 小结或知识扩展

response 对象的 sendRedirect 方法是在用户的浏览器端工作,Web 服务器要求浏览器重新发送一个到被定向页面的请求。在浏览器地址栏上会出现重定向页面的 URL,且为绝对路径。

forward 动作标记也可以实现页面的跳转,如<jsp:forward page="info.jsp"/>。但使用 forward 动作标记与 response 对象调用 sendRedirect 不同。对两者的比较如下：

- forward 为服务器端跳转,浏览器地址栏不变；sendRedirect 为客户端跳转,浏览器地址栏改变为新页面的 URL。
- 执行到 forward 动作标记出现处停止当前 JSP 页面的继续执行,而转向标记中 page 属性指定的页面；sendRedirect 所有代码执行完毕之后再跳转。
- 使用 forward,通过 request 请求信息能够保留在下一个页面；使用 sendRedirect 不能保留 request 请求信息。

forward 传递参数的格式如下：

```jsp
<jsp:forward page="info.jsp">
    <jsp:param name="no" value="001"/>
    <jsp:param name="age" value="18"/>
</jsp:forward>
```

response 对象的 sendRedirect 传递参数的方式如下：

response.sendRedirect("info.jsp? no=001&age=18") ;

(4) 任务 3 代码模板的参考答案

【代码】:response.sendRedirect("example3_2_3.jsp");

3.2.4 实践环节

编写 login.jsp、server.jsp 和 loginSuccess.jsp 3 个 JSP 页面，如果在页面 login.jsp 中输入正确的用户名"kazhafei"和正确的密码"aobama"，单击"登录"按钮后提交给页面 server.jsp。在 server.jsp 页面中进行登录验证：如果输入正确，提示"成功登录，3 秒钟后进入 loginSuccess.jsp 页面"；如果输入不正确，重定向到 login.jsp 页面。先运行 login.jsp 页面，页面运行效果如图 3.4(a)～图 3.4(c)所示。

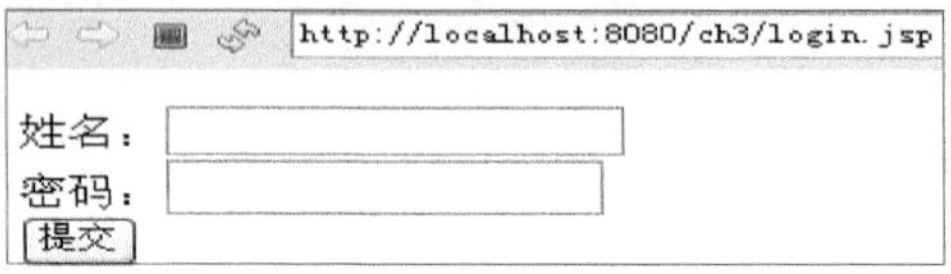

(a) login.jsp页面

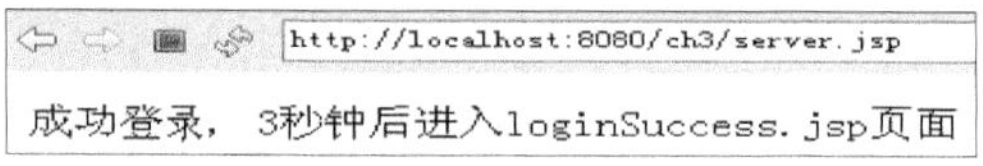

(b) server.jsp页面

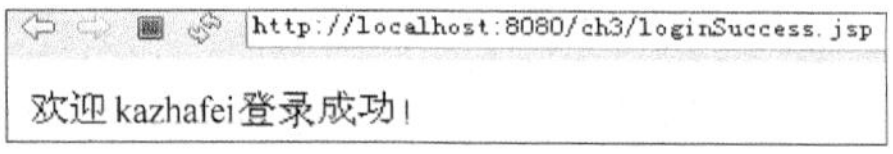

(c) loginSuccess.jsp页面

图 3.4　页面效果图

3.3　会话对象 session

浏览器与 Web 服务器之间使用 HTTP 协议进行通信。HTTP 是一种无状态协议，客户向服务器发出请求(request)，服务器返回响应(response)，连接就被关闭了，在服务器端不保留连接的相关信息。所以服务器必须采取某种手段来记录每个客户的连接信息。Web 服务器可以使用内置对象 session 来存放有关连接的信息。session 对象指的是客户端与服务器端的一次会话，从客户端连到服务器端的一个 Web 应用程序开始，直到客户端与服务器端断开为止。

3.3.1　核心知识

1. session 对象的 ID

Web 服务器会给每一个用户自动创建一个 session 对象，为每个 session 对象分配一个唯一标识的 String 类型的 session ID，这个 ID 用于区分其他用户。这样每个用户都对应着一个 session 对象(该用户的会话)，不同用户的 session 对象互不相同。session 对象调用 getId()方法就可以获取当前 session 对象的 ID。

2. session 对象存储数据

使用 session 对象可以保存用户在访问某个 Web 服务目录期间的有关数据。处理数据的方法如下：

- public void setAttribute(String key，Object obj)　将参数 obj 指定的对象保存到 session 对象中，key 为所保存的对象指定一个关键字。若保存的两个对象关键字相同，则先保存的对象被清除。
- public Object getAttribute(String key)　获取 session 中关键字是 key 的对象。
- public void removeAttribute(String key)　从 session 中删除关键字 key 所对应的对象。
- public Enumeration getAttributeNames()　产生一个枚举对象，该枚举对象可使用方法 nextElemets()遍历 session 中的各个对象所对应的关键字。

3. session 对象的生存期限

一个用户在某个 Web 服务目录中的 session 对象的生存期限依赖于以下几个因素。

- 用户是否关闭浏览器。
- session 对象是否调用 invalidate()方法。
- session 对象是否达到设置的最长"发呆"时间。

3.3.2　能力目标

理解 session 对象的生存期限，灵活使用 session 对象存储数据。

3.3.3　任务驱动

本节有以下 3 个任务。

1. 任务 1——获取 session 对象的 ID

(1) 任务 1 的主要内容

编写 example3_3_1.jsp、example3_3_2.jsp 和 example3_3_3.jsp 3 个 JSP 页面，其中，example3_3_2.jsp 存放在目录 tom 中，example3_3_3.jsp 存放在目录 cat 中。客户首先访问 example3_3_1.jsp 页面，从该页面链接到 example3_3_2.jsp 页面，然后从 example3_3_2.jsp 页面链接到 example3_3_3.jsp，效果如图 3.5(a)~图 3.5(c)所示。

(2) 任务 1 的代码模板

将下列 3 个 JSP 文件中的【代码】替换为真正的 JSP 的代码。

example3_3_1.jsp

```
<%@page language="java" contentType="text/html; charset=GBK" pageEncoding="GBK"%>
<html>
<head>
<title>example3_3_1.jsp</title>
</head>
<body>
    年轻人如何养生呢？<br><br>
```

(a) example3_3_1.jsp页面中的session对象的ID 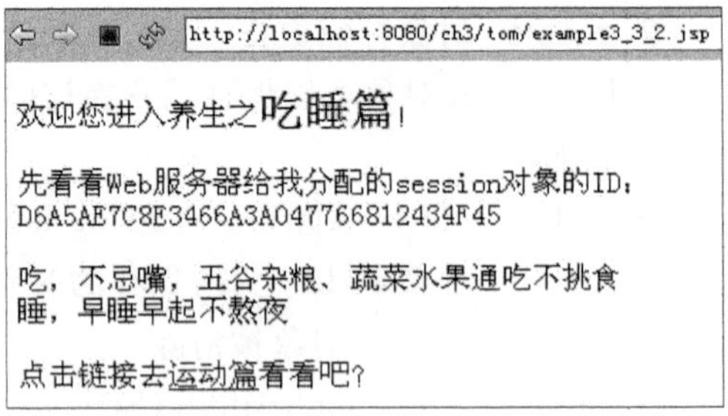   (b) example3_3_2.jsp页面中的session对象的ID

(c) example3_3_3.jsp页面中的session对象的ID

图 3.5　获取 session 对象的 ID

先看看 Web 服务器给我分配的 session 对象的 ID：
```
<%
    String id=【代码 1】           //使用session对象调用getId方法获得ID
%>
<br>
<%=id %>
<br><br>
点击链接去<a href="tom/example3_3_2.jsp">吃睡篇</a>看看吧？
</body>
</html>
```

example3_3_2.jsp

```
<%@page language="java" contentType="text/html; charset=GBK" pageEncoding="GBK"%>
<html>
<head>
<title>example3_3_2.jsp</title>
</head>
<body>
欢迎您进入养生之<font size=5>吃睡篇</font>!<br><br>
先看看 Web 服务器给我分配的 session 对象的 ID：
<%
    String id=【代码 2】           //使用session对象调用getId方法获得ID
%>
<br>
<%=id %>
<br><br>
```

吃,不忌嘴,五谷杂粮、蔬菜水果通吃不挑食`<br>`
睡,早睡早起不熬夜`<br><br>`
点击链接去`<a href="../cat/example3_3_3.jsp">`运动篇`</a>`看看吧?
`</body>`
`</html>`

example3_3_3.jsp

```
<%@page language="java" contentType="text/html; charset=GBK" pageEncoding="GBK"%>
<html>
<head>
<title>example3_3_3.jsp</title>
</head>
<body>
```
欢迎您进入养生之`<font size=5>`运动篇`</font>`!`<br><br>`
先看看Web服务器给我分配的session对象的ID:
```
<%
    String id=【代码3】            //使用session对象调用getId方法获得ID
%>
<br>
<%=id %>
<br><br>
```
动,坚持运动——这一点年轻人很多都做不好,`<br>`高兴起来就拼命打球,懒起来拼命睡觉,不好!`<br>`
总之,生活规律化,坚持长期运动`<br><br>`
点击链接去`<a href="../example3_3_1.jsp">`首页`</a>`看看吧?
`</body>`
`</html>`

(3)任务1小结或知识扩展

从任务1各个页面的运行结果看,一个用户在同一个Web服务目录中只有一个session对象,当用户访问相同Web服务目录的其他页面时,Web服务器不会再重新分配session对象,直到用户关闭浏览器或这个session对象达到了它的生存期限。当用户重新打开浏览器再访问该Web服务目录时,Web服务器为该客户再创建一个新的session对象。

需要注意的是,同一用户在多个不同的Web服务目录中所对应的session对象是不同的,一个服务目录对应一个session对象。

(4)任务1代码模板的参考答案

【代码1】:session.getId();
【代码2】:session.getId();
【代码3】:session.getId();

2. 任务2——使用session对象存储数据

(1)任务2的主要内容

使用session对象模拟在线考试系统。编写example3_3_4.jsp、example3_3_5.jsp和example3_3_6.jsp 3个JSP页面,在example3_3_4.jsp页面中考试,在example3_3_5.jsp页面中

显示答题结果,在 example3_3_6.jsp 页面中计算并公布考试成绩。先运行 example3_3_4.jsp 页面,效果如图 3.6(a)～图 3.6(c)所示。

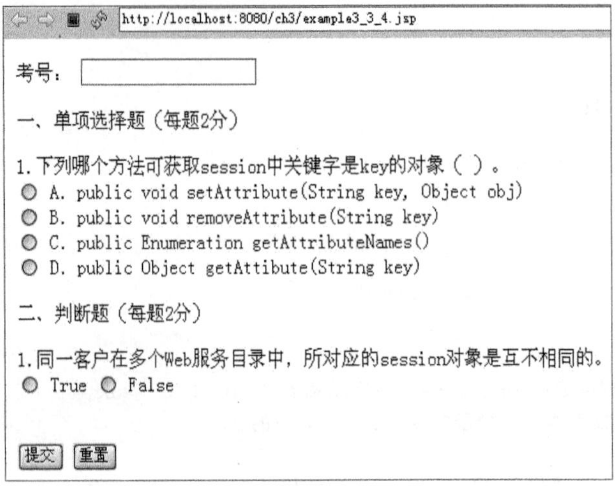

(a) 试卷页面

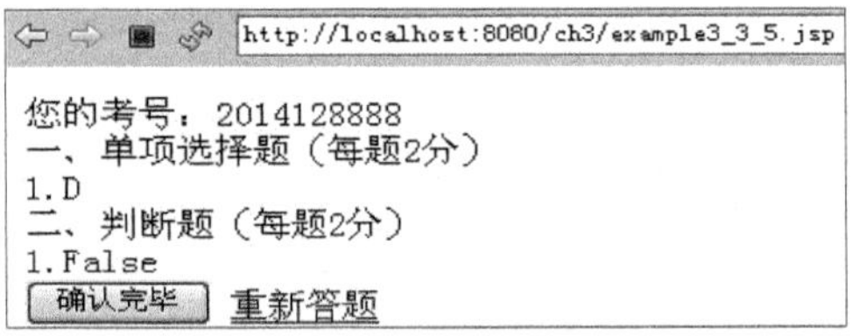

(b) 确认页面

(c) 成绩公布页面

图 3.6　session 模拟考试系统

(2) 任务 2 的代码模板

将下列 example3_3_5.jsp、example3_3_6.jsp 中的【代码】替换为真正的 JSP 的代码。

example3_3_4.jsp

```
<%@page language="java" contentType="text/html; charset=GBK" pageEncoding="GBK"%>
<html>
<head>
<title>example3_3_4.jsp</title>
</head>
<body>
    <form action="example3_3_5.jsp" method="post">
        考号:
        <input type="text" name="id"/>
        <p>
        一、单项选择题(每题 2 分)
        <br/><br/>
        1. 下列哪个方法可获取 session 中关键字是 key 的对象( )。
        <br />
        <input type="radio" name="one" value="A"/>
        A. public void setAttribute(String key, Object obj)<br/>
        <input type="radio" name="one" value="B"/>
```

```
                B. public void removeAttribute(String key)<br/>
                <input type="radio" name="one" value="C"/>
                C. public Enumeration getAttributeNames()<br/>
                <input type="radio" name="one" value="D"/>
                D. public Object getAttibute(String key)<br/>
            </p>
            <p>
                二、判断题(每题2分)
                <br/><br/>
                1.同一客户在多个Web服务目录中,所对应的session对象是互不相同的。
                <br/>
                <input type="radio" name="two" value="True"/>
                True
                <input type="radio" name="two" value="False"/>
                False
            </p><br/>
            <input type="submit" value="提交" name=submit>
            <input type="reset" value="重置" name=reset>
        </form>
</body>
</html>
```

example3_3_5.jsp

```
<%@page language="java" contentType="text/html; charset=GBK" pageEncoding="GBK"%>
<html>
<head>
<title>example3_3_5.jsp</title>
</head>
<body>
    <form action="example3_3_6.jsp" method="post">
            <%
                //考号
                String id=request.getParameter("id");
                【代码1】       //把考号id以"id"为关键字存储到session对象中
                //单项选择第一题
                String first=request.getParameter("one");
                【代码2】       //把答案first以"one"为关键字存储到session对象中
                //判断第一题
                String second=request.getParameter("two");
                【代码3】       //把答案second以"two"为关键字存储到session对象中
            %>
            您的考号:<%=id%><br/>
            一、单项选择题(每题2分)
            <br/>
            1.<%=first%>
            <br />
            二、判断题(每题2分)
            <br />
            1.<%=second%><br/>
```

```
                    <input type="submit" value="确认完毕"/>
                    <a href="example3_3_4.jsp">重新答题</a>
            </form>
</body>
</html>
```

example3_3_6.jsp

```
<%@page language="java" contentType="text/html; charset=GBK" pageEncoding="GBK"%>
<html>
<head>
<title>example3_3_6.jsp</title>
</head>
<body>
    <%
            //获取考号
            String id=(String)【代码 4】      //获取 session 中关键字是 id 的对象(考号)
            //计算成绩
            int sum=0;
            //如果单项选择第一题选中 D 选项,得 2 分
            String first=(String)【代码 5】
                                        //获取 session 中关键字是 one 的对象(选择答案)
            if ("D".equals(first)) {
                sum +=2;
            }
            //如果判断第一题选中 True,得 2 分
            String second=(String)【代码 6】
                                        //获取 session 中关键字是 two 的对象(判断答案)
            if ("True".equals(second)) {
                sum +=2;
            }
    %>
        您的成绩公布如下:
        <table border="1">
            <tr>
                <th width="50%">
                    考号
                </th>
                <th width="50%">
                    成绩
                </th>
            </tr>
            <tr>
                <td><%=id%></td>
                <td align="right"><%=sum%></td>
            </tr>
        </table>
</body>
</html>
```

(3) 任务 2 小结或知识扩展

从任务 2 中可以看出，一个用户进入某个 Web 服务目录，服务器为他分配一个 session 对象；用户可在该 Web 服务目录的所有页面使用 session 对象，该 session 可以存储、获取和移除用户的信息对象；当用户离开该 Web 服务目录时，session 对象就消失。

(4) 任务 2 代码模板的参考答案

【代码 1】:session.setAttribute("id", id);
【代码 2】:session.setAttribute("one", first);
【代码 3】:session.setAttribute("two", second);
【代码 4】:session.getAttribute("id");
【代码 5】:session.getAttribute("one");
【代码 6】:session.getAttribute("two");

3. 任务 3——session 对象的生存期限

(1) 任务 3 的主要内容

编写一个 JSP 页面 example3_3_7.jsp。如果用户是第一次访问该页面，会显示欢迎信息，并输出会话对象允许的最长"发呆"时间、创建时间，以及 session 的 ID。在 example3_3_7.jsp 页面中，session 对象使用 setMaxInactiveInterval(int maxValue) 方法设置最长的"发呆"状态时间为 10 秒。用户如果两次刷新间隔时间超过 10 秒，用户先前的 session 被取消，用户将获得一个新的 session 对象。页面运行效果如图 3.7(a)、图 3.7(b)所示。

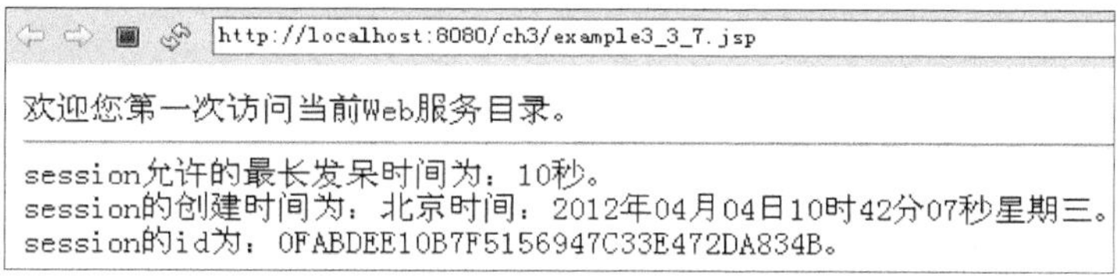

(a) 第一次或间隔10秒后访问该页面

(b) 10秒之内访问该页面

图 3.7　session 生存期限

(2) 任务 3 的代码模板

将下列 example3_3_7.jsp 中的【代码】替换为真正的 JSP 的代码。

example3_3_7.jsp

```
<%@ page language="java" contentType="text/html; charset=GBK" pageEncoding="GBK"%>
<%@ page import="java.util.*"%>
<%@ page import="java.text.*"%>
<html>
<head>
<meta http-equiv="Content-Type" content="text/html; charset=ISO-8859-1">
<title>example3_3_7.jsp</title>
</head>
```

```
<body>
    <%
【代码1】
        //session 调用 setMaxInactiveInterval(int n)方法设置最长"发呆"时间为10秒
boolean flg =【代码2】        //session 调用 isNew()方法判断 session 是不是新创建的
if (flg) {
    out.println("欢迎您第一次访问当前 Web 服务目录。");
    out.println("<hr/>");
}
out.println("session 允许的最长发呆时间为:" +
    session.getMaxInactiveInterval()+"秒。");
//获取 session 对象被创建的时间
long num=session.getCreationTime();
//将整数转换为 Date 对象
Date time=new Date(num);
//用给定的模式和默认语言环境的日期格式符号构造 SimpleDateFormat 对象
SimpleDateFormat matter=new SimpleDateFormat(
        "北京时间:yyyy 年 MM 月 dd 日 HH 时 mm 分 ss 秒 E。");
//得到格式化后的字符串
String strTime=matter.format(time);
out.println("<br/>session 的创建时间为:" +strTime);
out.println("<br/>session 的 id 为:" +session.getId() +"。");
%>
</body>
</html>
```

(3) 任务3小结或知识扩展

从任务3中可以看出，如果用户长时间不关闭浏览器，session 对象也没有调用 invalidate()方法，那么用户的 session 也可能消失。例如该任务中的 JSP 页面在10秒之内不被访问的话，它先前创建的 session 对象就消失了，服务器又重新创建一个 session 对象。这是因为 session 对象达到了它的最大"发呆"状态时间。所谓"发呆"状态时间，是指用户对某个 Web 服务目录发出的两次请求之间的间隔时间。

用户对某个 Web 服务目录下的 JSP 页面发出请求并得到响应，如果用户不再对该 Web 服务目录发出请求，比如不再操作浏览器，那么用户对该 Web 服务目录进入"发呆"状态，直到用户再次请求该 Web 服务目录时，"发呆"状态结束。

Tomcat 服务器允许用户最长的"发呆"状态时间为30分钟。可以通过修改 Tomcat 安装目录中 conf 文件夹下的配置文件 web.xml 文件，找到下面的片段，修改其中的默认值"30"，就可以重新设置各个 Web 服务目录下的 session 对象的最长"发呆"状态时间。这里的时间单位为分钟。

```
<session-config>
    <session-timeout>30</session-timeout>
</session-config>
```

也可以通过 session 对象调用 setMaxInactiveInterval(int time)方法来设置最长"发呆"状态时间，参数的时间单位为秒。

（4）任务3代码模板的参考答案

【代码1】：`session.setMaxInactiveInterval(10);`
【代码2】：`session.isNew();`

3.3.4 实践环节

用户到便民超市采购商品，购物前需要先登录会员卡号，购物时先把选购的商品放入购物车，最后到柜台清点商品。请借助于session对象模拟购物车，并存储客户的会员卡号和购买的商品名称。会员卡号输入后可以修改，购物车中的商品可以查看。编写程序模拟上述过程。loginID.jsp实现会员卡号输入，shop.jsp实现商品导购，food.jsp实现商品购物，count.jsp实现清点商品。本节实践环节的4个JSP页面都保存在目录shijian3中，先运行loginID.jsp页面，效果如图3.8(a)～图3.8(d)所示。

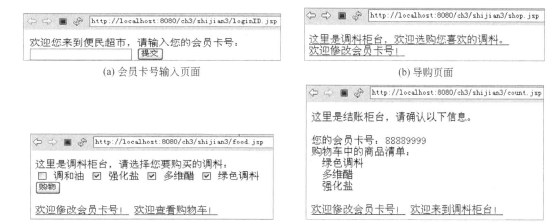

图 3.8　session模拟购物车

3.4　全局应用程序对象 application

3.4.1 核心知识

不同用户的session对象互不相同，但有时候用户之间可能需要共享一个对象，Web服务器启动后，就产生了这样一个唯一的内置对象application。任何用户在访问同一Web服务目录的各个页面时，共享一个application对象，直到服务器关闭，这个application对象被取消为止。application对象同session对象一样也可以进行数据的存储，处理数据的方法如下：

- public void setAttribute(String key，Object obj)　将参数obj指定的对象保存到application对象中，key为所保存的对象指定一个关键字。若保存的两个对象关键字相同，则先保存的对象被清除。
- public Object getAttribute(String key)　获取application中关键字是key的对象。
- public void removeAttribute(String key)　从application中删除关键字key所对应的对象。

- public Enumeration getAttributeNames()　产生一个枚举对象,该枚举对象可使用方法 nextElemets()遍历 application 中的各个对象所对应的关键字。

3.4.2　能力目标

理解 application 对象的生存期限,灵活使用 application 对象存储数据。

3.4.3　任务驱动

1. 任务的主要内容

用 application 制作"四字成语接龙",用户通过 example3_4_1.jsp 页面向 example3_4_2.jsp 页面提交四字成语,example3_4_2.jsp 页面获取成语内容后,用同步方法将该成语内容和以前的成语内容进行连接,然后将这些四字成语内容添加到 application 对象中。页面运行效果如图 3.9(a)、图 3.9(b)所示。

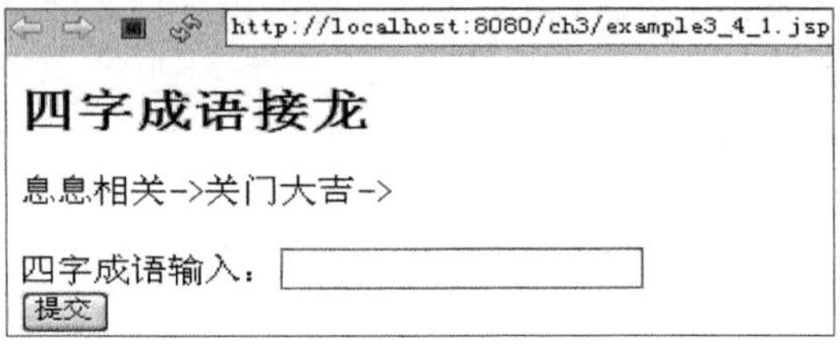

(a) 成语提交页面

(b) 接龙成功页面

图 3.9　成语接龙

2. 任务的代码模板

将下列两个 JSP 文件中的【代码】替换为真正的 JSP 的代码。

example3_4_1.jsp

```
<%@page language="java" contentType="text/html; charset=GBK" pageEncoding="GBK"%>
<html>
<head>
<title>example3_4_1.jsp</title>
</head>
<body>
<h2>四字成语接龙</h2>
<%
    String s=【代码 1】     //取出 application 中关键字是 message 的对象(成语内容)
    if(s!=null){
        out.print(s);
    }
    else{
        out.print("还没有词语,请您龙头开始!<br>");
    }
```

```
%>
<form action="example3_4_2.jsp" method="post">
    四字成语输入:<input type="text" name="mes"/><br>
    <input type="submit" value="提交"/>
</form>
</body>
</html>
```

example3_4_2.jsp

```
<%@page language="java" contentType="text/html; charset=GBK" pageEncoding="GBK"%>
<%@page import="java.util.*" %>
<html>
<head>
<title>example3_4_2.jsp</title>
</head>
<body>
    <%!
    String message="";
    ServletContext application;
    synchronized void sendMessage(String s){
        application=getServletContext();
        message=message+s+"->";
        【代码2】  //把成语内容message以"message"为关键字存储到application对象中
    }
    %>
    <%
    String content=request.getParameter("mes");
    byte b[]=content.getBytes("ISO-8859-1");
    content=new String(b);
    sendMessage(content);
    out.print("您的四字成语已经提交!3秒钟后回到成语页面,继续接龙!");
    response.setHeader("refresh", "3;url=example3_4_1.jsp");
    %>
</body>
</html>
```

3. 任务小结或知识扩展

任务中成语接龙方法sendMessage()为什么定义为同步方法呢？这是因为application对象对所有的用户都是相同的,任何用户对该对象中存储的数据的操作都会影响到其他用户。

如果客户浏览不同的Web服务目录,将产生不同的application对象。同一个Web服务目录中的所有JSP页面都共享同一个application对象,即使浏览这些JSP页面的是不同的客户也是如此。因此,保存在application对象中的数据不仅可以跨页面分享,还可以由所有用户共享。

有些Web服务器不能直接使用application对象,必须使用父类ServletContext声明这个对象,然后使用getServletContext()方法为application对象进行实例化。例如该任务的

example3_4_2.jsp 页面中的代码。

4. 代码模板的参考答案

【代码 1】:(String)application.getAttribute("message");
【代码 2】:application.setAttribute("message", message);

3.4.4 实践环节

使用 application 对象实现网站访客计数器的功能。计数器运行效果如图 3.10 所示。

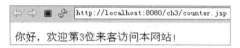

图 3.10 计数器页面

3.5 小　　结

- 所有的内置对象不需要由 JSP 的编写者声明和实例化，可以直接在所有的 JSP 网页中使用。内置对象只在 Java 程序片或者 Java 表达式中使用。
- 请求对象 request 代表客户端发出的请求信息对象,通常用来获取表单上的信息。request 对象只在从客户发出请求到服务器作出响应此期间是有效的。
- 响应对象 response 代表从服务器端返回给客户端的响应信息对象。response 对象包含从动态页面返回给客户的所有信息。response 对象可以用来进行页面的重定向,但与用 forward 动作标记实现页面的跳转有所不同。
- 会话对象 session 可以用来保存每个用户信息,以便跟踪每个用户的操作状态,不同用户的 session 对象互不相同。session 对象主要用来存储和获取数据,使用时要注意其生存期限。
- 全局应用程序对象 application 显示相应网页所用应用程序的对象。任何用户在访问同一 Web 服务目录的各个页面时,共享一个 application 对象,直到服务器关闭,这个 application 对象被取消为止。保存在 application 对象中的数据不仅可以跨页面分享,还可以由所有用户共享。

习　题　3

1. 下面(　　)操作不能关闭 session 对象。
   A. 用户刷新当前页面调用　　　　　　B. 用户关闭浏览器
   C. session 达到设置的最长"发呆"状态时间　　D. session 对象的 invalidate()方法
2. 有如下程序片段。

```
<form>
<input type="text" name="id">
<input type="submit" value ="提交">
```

</form>

下面(　　)语句可以获取用户输入的信息。

A. request.getParameter("id");

B. request.getAttribute("submit");

C. session.getParameter(key,"id");

D. session.getAttribute(key,"id");

3. 下面(　　)内置对象是对客户的请求作出响应,向客户端发送数据的。

A. request　　　B. session　　　C. response　　　D. application

4. 可以使用(　　)方法实现客户的重定向。

A. response.setStatus();　　　　B. response.setHeader();

C. response.setContentType();　　D. response.sendRedirect();

5. 什么对象是内置对象？常见的内置对象有哪些？

6. 请简述内置对象 request、session 和 application 之间的区别。

7. 一个用户在不同 Web 服务目录中的 session 对象相同吗？一个用户在同一 Web 服务目录的不同子目录中的 session 对象相同吗？

8. session 对象的生存期限依赖于哪些因素？

9. 简述 forward 动作标记与 response.sendRedirect() 两种跳转的区别。

JSP 与 JavaBean

本章主要内容

- 编写和使用 JavaBean
- 获取和修改 bean 的属性

我们已经知道一个 JSP 页面通过使用 HTML 标记为用户显示数据（静态部分），页面中变量的声明、程序片以及表达式为动态部分，对数据进行处理。如果 Java 程序片和 HTML 标记大量掺杂在一起使用，就不利于 JSP 页面的扩展和维护。JSP 和 JavaBean 技术的结合不仅可以实现数据的表示和处理分离，而且可以提高 JSP 程序代码重用的程度，是 JSP 编程中常用的技术。

在本章中，将新建一个 Web 工程 ch4，本章例子中涉及的 Java 源文件保存在 ch4 的 src 中，涉及的 JSP 页面保存在 ch4 的 WebContent 目录中。

4.1 编写 JavaBean

 核心知识

JavaBean 是一个可重复使用的软件组件，是遵循一定标准、用 Java 语言编写的一个类，该类的一个实例称为一个 JavaBean，简称 bean。

编写一个 JavaBean 就是编写一个 Java 的类（该类必须带有包名），这个类创建的一个对象称为一个 bean，为了让 JSP 引擎（比如 Tomcat）知道这个 bean 的属性和方法，必须在类的方法命名上遵守以下规则。

（1）如果类的成员变量的名字是 name，那么为了获取或更改成员变量的值，类中必须提供如下两个方法。

- getName()，用来获取属性 name。
- setName()，用来修改属性 name。

即方法的名字用 get 或 set 为前缀，后缀是首字母大写的成员变量的名字。

（2）对于 boolean 类型的成员变量，允许使用"is"代替上面的"get"和"set"。

（3）类中声明的方法的访问权限都必须是 public。

（4）类中声明的构造方法必须是 public、无参数。

4.1.2 能力目标

能够灵活使用 JavaBean 的编写规则编写创建 bean 的 Java 源文件。

4.1.3 任务驱动

1. 任务的主要内容

创建 bean 的源文件 Rectangle.java(在包 small.dog 中)，该 bean 的作用是计算矩形的面积和周长。

2. 任务的代码模板

将下列 Rectangle.java 中的【代码】替换为真正的 Java 代码。

Rectangle.java

```
package small.dog;
public class Rectangle {
    double length;
    double width;
    【代码 1】{                               //定义类 Rectangle 的构造方法
        length=10;
        width=5;
    }
    【代码 2】{                               //定义获取矩形长度的方法
        return length;
    }
    【代码 3】{                               //定义修改矩形长度的方法
        this.length=length;
    }
    public double getWidth() {
        return width;
    }
    public void setWidth(double width) {
        this.width=width;
    }
    public double computerArea(){
        return length* width;
    }
    public double computerLength(){
        return (length+width)* 2;
    }
}
```

3. 任务小结或知识扩展

JavaBean 可以在任何 Java 程序编写环境下完成编写，再通过编译成为一个字节码文件(.class 文件)，为了让 JSP 引擎(比如 Tomcat)找到这个字节码，必须把字节码文件放在特定的位置。本书使用 Eclipse 集成环境开发 JSP 程序，Java 类的字节码文件由 Eclipse 自动保存到 Web 工程的 build\classes 中。比如，该任务中的 Rectangle.class 文件保存在 ch4\build\classes\small\dog 目录中。

我们知道JavaBean是基于Java语言的,因此JavaBean具有以下特点。
- 与平台无关。
- 代码的重复利用。
- 易扩展,易维护,易使用。

4. 代码模板的参考答案

【代码1】:`public Rectangle()`
【代码2】:`public double getLength()`
【代码3】:`public void setLength(double length)`

4.1.4 实践环节

创建bean的源文件Circle.java(在包big.dog中),该bean的作用是计算圆形的面积和周长。

4.2 JSP页面中创建与使用bean

4.2.1 核心知识

在JSP页面中使用bean,首先必须使用page指令的import属性导入创建bean的类所在的包,例如:

`<%@page import=" small.dog.* "%>`

其次使用JSP动作标记useBean,来创建与使用bean。useBean标记的格式为

`<jsp:useBean  id="bean 的名字" class="创建 bean 的类" scope="bean 的有效范围"/>`

或

```
<jsp:useBean  id="bean 的名字" class="创建 bean 的类" scope="bean 的有效范围">
</jsp:useBean>
```

例如:

`<jsp:useBean  id="rectangle" class="small.dog.Rectangle" scope="page"/>`

当含有useBean动作标记的JSP页面被JSP引擎(比如Tomcat)加载执行时,JSP引擎首先根据id的名字,在pageContent内置对象中查看是否含有名字id和作用域scope的对象;如果该对象存在,JSP引擎就将这个对象的副本(bean)分配给JSP页面使用;如果没有找到,就根据class指定的类创建一个名字是id的bean,并添加到pageContent对象中,同时将这个bean分配给JSP页面使用。useBean动作标记执行流程如图4.1所示。

4.2.2 能力目标

在JSP页面中能够灵活使用动作标记useBean。

第 4 章 JSP 与 JavaBean

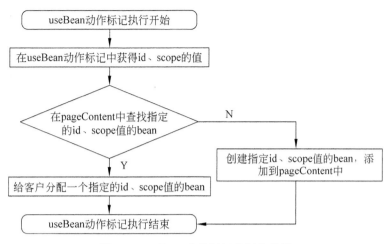

图 4.1　useBean 动作标记执行流程图

4.2.3　任务驱动

1. 任务的主要内容

编写一个 JSP 页面 example4_1.jsp, 在 JSP 页面中使用 useBean 动作标记获得一个 bean, 负责创建 bean 的类是 4.1 节任务中的 Rectangle 类, 创建 bean 的名字是 rectangle, rectangle 的 scope 取值为 page。JSP 页面的运行效果如图 4.2 所示。

图 4.2　使用 bean 的 JSP 页面

2. 任务的代码模板

将下列 example4_1.jsp 中的【代码】替换为真正的 JSP 的代码。

example4_1.jsp

```
<%@page language="java" contentType="text/html; charset=GBK" pageEncoding=
"GBK"%>
<%@page import="small.dog.Rectangle"%>
<html>
<head>
<title>example4_1.jsp</title>
</head>
<body>
    【代码】<%--通过 useBean 动作标记获得一个 bean,负责创建 bean 的类是 small.dog.
    Rectangle,id 是 rectangle,scope 取值为 page --%>
    <p>矩形的长是:<%=rectangle.getLength()%>
```

```
            <p>矩形的宽是:<%=rectangle.getWidth() %>
            <p>矩形的面积是:<%=rectangle.computerArea() %>
            <p>矩形的周长是:<%=rectangle.computerLength() %>
    </body>
</html>
```

3. 任务小结或知识扩展

从创建 bean 的过程可以看出,首次创建一个新的 bean 需要用相应的字节码文件创建对象,当别的 JSP 页面再需要同样的 bean 时,JSP 引擎直接将 pageContent 内置对象里已经存在的对象的副本分配给相应的 JSP 页面,提高了代码的复用程度。如果程序员修改了字节码文件,必须重启 JSP 引擎,才能使用新的字节码文件。

useBean 动作标记中 scope 的默认值是 page,除 page 之外,scope 的取值还有 request、session 与 application。

(1) scope 取值 page

该 bean 的有效范围是当前页面。当客户请求 bean 时,分配内存空间给它;当客户离开这个页面时,便取消分配的 bean,并收回内存空间。JSP 引擎分配给每个 JSP 页面的 bean 是不同的,它们占有不同的内存空间。

当两个客户访问同一个 JSP 页面时,一个用户对自己 bean 的属性的改变,不会影响到另一个客户。

(2) scope 取值 request

该 bean 的有效范围是 request 期间。客户在网站访问时请求多个页面,如果每个页面都含有 useBean 动作标记,那么在每个页面分配的 bean 也不相同。JSP 引擎对请求作出响应后,bean 消失。

当两个客户同时请求一个 JSP 页面时,一个用户对自己 bean 属性的改变,不会影响另外一个客户。

(3) scope 取值 session

该 bean 的有效范围是客户的会话期间。如果客户在多个页面中互相连接,每个页面都含有一个相同的 useBean 动作标记,那么这个客户在这些页面得到的 bean 是相同的,即占有相同的内存空间。当会话结束时,bean 消失,释放空间。

如果一个客户在某个页面更改了 bean 的某个属性,那么该客户的其他页面 bean 的属性也发生变化。但两个客户同时访问一个 JSP 页面时,一个客户对自己 bean 的属性的改变不会影响到另一个客户。

(4) scope 取值 application

该 bean 的有效范围是 application 期间(Web 服务器启动期间)。JSP 引擎为所有的 JSP 页面分配一个共享的 bean。

当几个客户同时访问一个 JSP 页面时,任何一个客户对自己 bean 的属性的改变都会影响到其他客户。

4. 代码模板的参考答案

【代码】:<jsp:useBean id="rectangle" class="small.dog.Rectangle" scope="page"/>

4.2.4 实践环节

编写一个 JSP 页面 computerCircle.jsp,在 JSP 页面中使用 useBean 动作标记获得一个 bean,负责创建 bean 的类是 4.1.4 小节实践环节中的 Circle 类,创建 bean 的名字为 circle,circle 的 scope 取值为 request。JSP 页面的运行效果如图 4.3 所示。

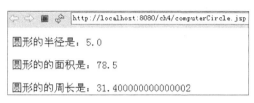

图 4.3 使用 bean 计算圆形的面积与周长

4.3 获取 bean 的属性

JavaBean 的实质是遵守一定规范的类所创建的对象,可以通过如下两种方式获取 bean 的属性。

1. Java 程序片

可以通过调用构造方法获得一个 bean,然后调用 getXXX()方法来获取 bean 的属性。

2. JSP 标记

先通过<jsp:useBean>标记获得一个 bean,再通过<jsp:getProperty>标记获取 bean 的属性值(无须使用 Java 程序片)。

4.3.1 核心知识

使用 getProperty 动作标记可以获得 bean 的属性值。使用该动作标记之前,必须事先使用 useBean 动作标记获得一个相应的 bean。getProperty 动作标记语法格式如下:

`<jsp:getProperty  name="bean 的名字"  property="bean 的属性" />`

或

`< jsp: getProperty  name = " bean 的名字"  property = " bean 的属性"/> </jsp: getProperty>`

其中,name 取值是 bean 的名字,和 useBean 动作标记中的 id 对应;property 取值是 bean 的一个属性的名字,和创建该 bean 的类的成员变量名对应。这条指令相当于在 Java 表达式中使用 bean 的名字调用 getXXX()方法。

4.3.2 能力目标

能够灵活使用 getProperty 动作标记获得 bean 的属性。

4.3.3 任务驱动

1. 任务的主要内容

- 创建 bean 的源文件 NewRectangle.java（在包 small.dog 中），该 bean 的作用是计算矩形的面积和周长。
- 编写一个 JSP 页面 useGetProperty.jsp，在该 JSP 页面中使用 useBean 动作标记创建一个名字是 pig 的 bean，并使用 getProperty 动作标记获得 pig 的每个属性的值。负责创建 pig 的类是 NewRectangle 类。JSP 页面运行效果如图 4.4 所示。

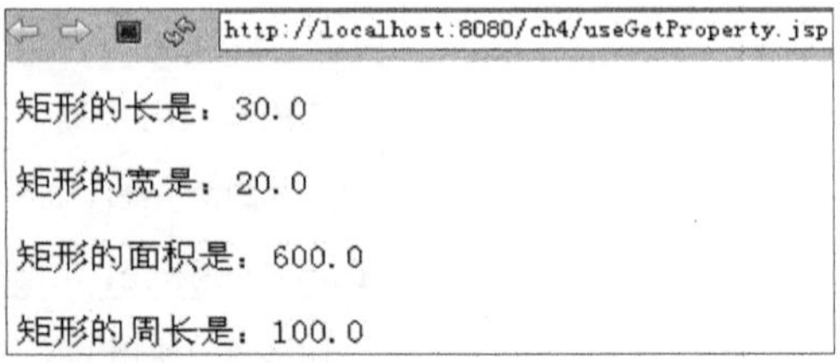

图 4.4 使用 bean 计算矩形的面积与周长

2. 任务的代码模板

将下列 useGetProperty.jsp 中的【代码】替换为真正的 JSP 的代码。

NewRectangle.java

```java
package small.dog;
public class NewRectangle {
    double length;
    double width;
    double rectangleArea;
    double rectangleLength;
    public NewRectangle(){
        length=20;
        width=10;
    }
    public double getLength() {
        return length;
    }
    public void setLength(double length) {
        this.length=length;
    }
    public double getWidth() {
        return width;
    }
    public void setWidth(double width) {
        this.width=width;
    }
    public double getRectangleArea() {
        return length * width;
    }
    public double getRectangleLength() {
        return 2 * (width+length);
    }
}
```

useGetProperty.jsp

```jsp
<%@page language="java" contentType="text/html; charset=GBK" pageEncoding="GBK"%>
<%@page import="small.dog.NewRectangle"%>
<html>
<head>
<title>useGetProperty.jsp</title>
</head>
<body>
    <jsp:useBean id="pig" class="small.dog.NewRectangle" scope="page"/>
    <%pig.setLength(30);%>
    <%pig.setWidth(20);%>
    <p>矩形的长是：【代码 1】<%--使用 getProperty 动作标记获得矩形的长 --%>
    <p>矩形的宽是：【代码 2】<%--使用 getProperty 动作标记获得矩形的宽 --%>
    <p>矩形的面积是：【代码 3】<%--使用 getProperty 动作标记获得矩形的面积 --%>
    <p>矩形的周长是：【代码 4】<%--使用 getProperty 动作标记获得矩形的周长 --%>
</body>
</html>
```

3. 任务小结或知识扩展

在 JSP 页面中使用 getProperty 动作标记获得 bean 的属性时，必须保证 bean 中有相应的 getXXX()方法，即创建 bean 的类中定义了 getXXX()方法。

从 useGetProperty.jsp 页面可以看出，使用 getProperty 动作标记获得 bean 的属性值，减少了 Java 程序片的使用。

在 NewRectangle.java 中，有两个属性（rectangleArea 和 rectangleLength）没有提供 set()方法，因为两者依赖于 length 和 width 属性。

4. 代码模板的参考答案

【代码 1】：`<jsp:getProperty property="length" name="pig"/>`
【代码 2】：`<jsp:getProperty property="width" name="pig"/>`
【代码 3】：`<jsp:getProperty property="rectangleArea" name="pig"/>`
【代码 4】：`<jsp:getProperty property="rectangleLength" name="pig"/>`

4.3.4 实践环节

（1）创建 bean 的源文件 Ladder.java（在包 big.dog 中），该 bean 的作用是计算梯形的面积。

（2）编写一个 JSP 页面 ladderProperty.jsp，在该 JSP 页面中使用 useBean 动作标记创建一个名字是 lad 的 bean，并使用 getProperty 动作标记获得 lad 的每个属性的值。负责创建 lad 的类是 Ladder 类。JSP 页面运行效果如图 4.5 所示。

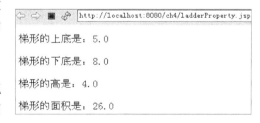

图 4.5 使用 getProperty 动作标记获得梯形的属性值

4.4 修改 bean 的属性

4.4.1 核心知识

使用 setProperty 动作标记可以修改 bean 的属性值。使用该动作标记之前,必须事先使用 useBean 动作标记获得一个相应的 bean。使用 setProperty 动作标记进行 bean 属性的设置有如下 3 种方式。

1. 用表达式或字符串设置 bean 的属性

(1) 用表达式设置 bean 的属性

`<jsp:setProperty name="bean 的名字" property="bean 的属性" value="<%=expression%>" />`

(2) 用字符串设置 bean 的属性

`<jsp:setProperty name="bean 的名字" property="bean 的属性" value=字符串 />`

2. 通过 HTTP 表单的参数的值设置 bean 的属性

`<jsp:setProperty name="bean 的名字" property="*" />`

3. 任意指定请求参数设置 bean 的属性

`<jsp:setProperty name="bean 的名字" property="属性名" param="参数名"/>`

可以根据自己的需要,任意选择传递的参数,请求参数名无须与 bean 属性名相同。

4.4.2 能力目标

能够灵活使用 setProperty 动作标记修改 bean 的属性。

4.4.3 任务驱动

本节有以下两个任务。

1. 任务 1——用表达式或字符串修改 bean 的属性

(1) 任务 1 的主要内容

- 创建 bean 的源文件 Car.java(在包 small.dog 中),该 bean 的作用是描述小汽车的一些属性。
- 编写一个 JSP 页面 car.jsp,在该 JSP 页面中使用 useBean 动作标记创建一个名字是 smallCar 的 bean,其有效范围是 page,并使用动作标记修改、获取该 bean 的属性的值。负责创建 smallCar 的类是 Car 类。JSP 页面运行效果如图 4.6 所示。

图 4.6 使用字符串或表达式的值修改 bean 的属性

(2) 任务 1 的代码模板

将下列 car.jsp 中的【代码】替换为真正的 JSP 的代码。

Car.java

```java
package small.dog;
public class Car {
    String tradeMark;
    String number;
    public String getTradeMark() {
        return tradeMark;
    }
    public void setTradeMark(String tradeMark) {
        this.tradeMark=tradeMark;
    }
    public String getNumber() {
        return number;
    }
    public void setNumber(String number) {
        this.number=number;
    }
}
```

car.jsp

```jsp
<%@page language="java" contentType="text/html; charset=GBK" pageEncoding="GBK"%>
<%@page import="small.dog.Car"%>
<html>
<head>
<title>car.jsp</title>
</head>
<body>
    <jsp:useBean id="smallCar" class="small.dog.Car" scope="page"/>
    <%
        String carNo="辽 B8888";
    %>
    【代码 1】<%--使用 setProperty 动作标记设置 smallCar 的属性 tradeMark 值为"宝马
        X6"--%>
    【代码 2】<%--使用 setProperty 动作标记设置 smallCar 的属性 number 值为 carNo--%>
    汽车的品牌是:<jsp:getProperty property="tradeMark" name="smallCar"/>
    <br>汽车的牌号是:<jsp:getProperty property="number" name="smallCar"/>
</body>
</html>
```

(3) 任务 1 小结或知识扩展

用表达式修改 bean 属性的值时,表达式值的类型必须与 bean 的属性类型一致。用字符串修改 bean 属性的值时,字符串会自动被转化为 bean 的属性的类型,不能转化成功的可能会抛出 NumberFormatException 异常。

(4) 任务 1 代码模板的参考答案

【代码 1】:`<jsp:setProperty property="tradeMark" name="smallCar" value="宝马 X6"/>`

【代码 2】：<jsp:setProperty property="number" name="smallCar" value="<%=carNo%>"/>

2. 任务 2——通过 HTTP 表单的参数的值设置 bean 的属性

(1) 任务 2 的主要内容

编写两个 JSP 页面：inputCar.jsp 和 showCar.jsp。在 inputCar.jsp 页面中输入信息后提交给 showCar.jsp 页面显示信息。页面中用到的 bean 是使用任务 1 中 Car 类创建的。页面运行效果如图 4.7(a)、图 4.7(b)所示。

(a) 信息输入页面　　　　　　　　　　　(b) 信息显示页面

图 4.7　任务 2 的页面运行效果图

(2) 任务 2 的代码模板

将下列 showCar.jsp 中的【代码】替换为真正的 JSP 的代码。

inputCar.jsp

```
<%@page language="java" contentType="text/html; charset=GBK" pageEncoding="GBK"%>
<html>
<head>
<title>inputCar.jsp</title>
</head>
<body>
    <form action="showCar.jsp" method="post" >
        请输入汽车品牌：
        <input type="text" name="tradeMark"/>
        <br>
        请输入汽车牌号：
        <input type="text" name="number"/>
        <br>
        <input type="submit" value="提交"/>
    </form>
</body>
</html>
```

showCar.jsp

```
<%@page language="java" contentType="text/html; charset=GBK" pageEncoding="GBK"%>
<%@page import="small.dog.Car"%>
<%
    request.setCharacterEncoding("GBK");
%>
<html>
```

```
<head>
<title>car.jsp</title>
</head>
<body>
    <jsp:useBean id="smallCar" class="small.dog.Car" scope="page"/>
    【代码】<%--通过 HTTP 表单的参数的值设置 bean 的属性(表单参数与属性自动匹配)--%>
    汽车的品牌是:<jsp:getProperty property="tradeMark" name="smallCar"/>
    <br>汽车的牌号是:<jsp:getProperty property="number" name="smallCar"/>
</body>
</html>
```

(3) 任务 2 小结或知识扩展

通过 HTTP 表单的参数的值设置 bean 的属性时,表单的参数的名字必须与 bean 属性的名字相同,服务器会根据名字自动匹配,类型自动转换。

由于客户可能通过表单提交汉字字符,可以采用 request.setCharacterEncoding("GBK") 避免出现中文乱码。采用该方式避免中文乱码时,表单的提交方式一定是 post 的方式。

(4) 任务 2 代码模板的参考答案

【代码】:<jsp:setProperty property="*" name="smallCar"/>

4.4.4 实践环节

编写两个 JSP 页面:inputNumber.jsp 与 showResult.jsp。inputNumber.jsp 提供一个表单,用户可以通过表单输入两个数和四则运算符号提交给 showResult.jsp。用户提交表单后,JSP 页面将计算任务交给一个 bean 去完成,创建 bean 的源文件 Computer.java(在包 big.dog 中)。页面运行效果如图 4.8(a)、图 4.8(b)所示。

(a) 数据输入页面　　　　　　　　　　(b) 结果显示页面

图 4.8　4.4 节实践环节页面运行效果图

4.5　JSP 与 bean 结合的简单例子

4.5.1 核心知识

通过前面的学习,我们已经知道在 JSP 页面中使用 JavaBean 可以将数据的处理代码从页面中分离出来,提高了代码的复用程度,方便了代码的维护。

为了进一步掌握 JavaBean 的应用,本节给出了一个具体的实际问题:登录验证。

4.5.2 能力目标

熟练掌握 JavaBean 的应用。

4.5.3 任务驱动

1. 任务的主要内容

编写两个 JSP 页面：login.jsp 与 invalidate.jsp。login.jsp 页面提供一个表单，用户通过表单将用户名和密码（正确的用户名和密码分别是 Obama 和 Qaddafi）提交给 invalidate.jsp 页面。用户提交表单后，JSP 页面将登录验证的任务提交给一个 bean 去完成。页面运行效果如图 4.9(a)、图 4.9(b)所示。

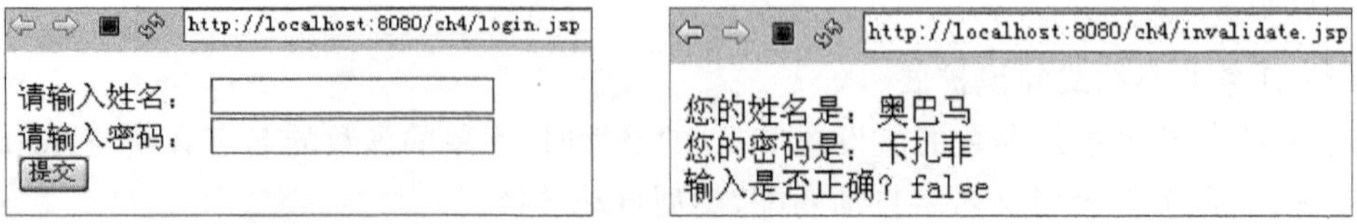

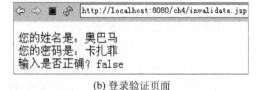

(a) 信息输入页面　　　　　　　　　　　(b) 登录验证页面

图 4.9　4.5 节任务的页面运行效果图

2. 任务的代码模板

将下列 login.jsp 中的【代码】替换为真正的 JSP 的代码。

LoginBean.java

```java
package small.dog;
public class LoginBean {
    String uname;
    String upass;
    boolean login;
    public String getUname() {
        return uname;
    }
    public void setUname(String uname) {
        this.uname=uname;
    }
    public String getUpass() {
        return upass;
    }
    public void setUpass(String upass) {
        this.upass=upass;
    }
    public boolean isLogin() {
        if(uname.equals("Obama")&&upass.equals("Qaddafi")){
            login=true;
        }else{
            login=false;
        }
        return login;
    }
    public void setLogin(boolean login) {
        this.login=login;
    }
```

}

login.jsp

```
<%@page language="java" contentType="text/html; charset=GBK" pageEncoding=
"GBK"%>
<html>
<head>
<title>login.jsp</title>
</head>
  <BODY>
  <FORM action="invalidate.jsp" method="post">
     请输入姓名：
       <Input type=text name="uname"/><BR>
     请输入密码：
       <Input type=text name="upass"/><BR>
     <INPUT TYPE="submit" value="提交"/>
  </FORM>
  </BODY>
</html>
```

invalidate.jsp

```
<%@page language="java" contentType="text/html; Charset=GBK" pageEncoding=
"GBK"%>
<%@page import="small.dog.LoginBean"%>
<%
request.setCharacterEncoding("GBK");
%>
<html>
<head>
<title>invalidate.jsp</title>
</head>
<body>
    <jsp:useBean id="lb" class="small.dog.LoginBean" scope="page"/>
    【代码1】<%--通过HTTP表单的参数的值设置bean的属性(表单参数与属性自动匹配)--%>
    您的姓名是:【代码2】<%--使用getProperty动作标记获得bean的属性值--%>
    <br>您的密码是:【代码3】<%--使用getProperty动作标记获得bean的属性值--%>
    <br>输入是否正确?【代码4】<%--使用getProperty动作标记获得bean的属性值--%>
</body>
</html>
```

3. 任务小结或知识扩展

如果bean的属性为boolean类型,可以使用isXXX()代替getXXX()方法,如本节任务中的LoginBean的login。

4. 代码模板的参考答案

【代码1】:`<jsp:setProperty property="*" name="lb"/>`
【代码2】:`<jsp:getProperty property="uname" name="lb"/>`
【代码3】:`<jsp:getProperty property="upass" name="lb"/>`

【代码 4】:`<jsp:getProperty property="login" name="lb"/>`

4.5.4 实践环节

编写两个 JSP 页面：inputTriangle.jsp 与 showTriangle.jsp。inputTriangle.jsp 提供一个表单，用户可以通过表单输入三角形的三条边提交给 showTriangle.jsp。用户提交表单后，JSP 页面将计算三角形的面积和周长的任务交给一个 bean 去完成，创建 bean 的源文件 Triangle.java（在包 big.dog 中）。页面运行效果如图 4.10(a)、图 4.10(b)所示。

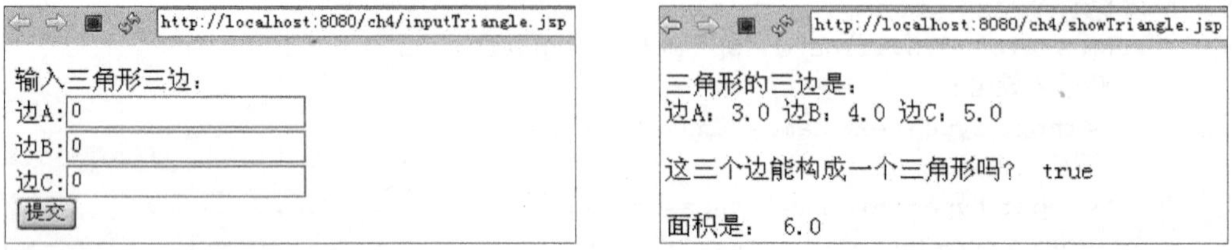

(a) 三角形边长输入页面　　　　　　　　(b) 信息显示页面

图 4.10　4.5 节实践环节页面运行效果图

4.6 小　　结

- JavaBean 的实质是一种特殊的 Java 类，是一个可重复使用的软件组件，是遵循一定标准、用 Java 语言编写的一个类，该类的一个实例称为一个 JavaBean，简称 bean。
- JSP 和 JavaBean 技术的结合不仅可以实现数据的表示和处理分离，而且提高了代码的可重用、可维护性。
- JavaBean 的生命期限分为 page、request、session 和 application。

习　题　4

1. 下面哪一个是正确使用 JavaBean 的方式？（　　）
   A. `<jsp:useBean id="address" class="tom.AddressBean" scope="page"/>`
   B. `<jsp:useBean name="address" class="tom.AddressBean" scope="page"/>`
   C. `<jsp:useBean bean="address" class="tom.AddressBean" scope="page" />`
   D. `<jsp:useBean beanName="address" class="AddressBean" scope="page" />`
2. JavaBean 中声明的方法的访问属性必须是（　　）。
   A. private　　　　B. public　　　　C. protected　　　　D. friendly
3. 在 JSP 中调用 JavaBean 时不会用到的标记是（　　）。
   A. `<javabean>`　　　　　　　　　　B. `<jsp:useBean>`
   C. `<jsp:setProperty>`　　　　　　D. `<jsp:getProperty>`
4. JavaBean 的作用域可以是（　　）、page、session 和 application。
   A. request　　　　B. response　　　　C. out　　　　D. 以上都不对
5. 在 test.jsp 文件中有如下一行代码。

```
<jsp:useBean class="tom.jiafei.Test" id="user" scope="____" />
```

要使 user 对象一直存在于会话中,直至终止或被删除为止,下画线中应填入(　　)。

    A. page   B. request   C. session   D. application

6. 关于 JavaBean 正确的说法是(　　)。

    A. 类中声明的方法的访问权限必须是 public

    B. 在 JSP 文件中引用 bean,其实就是用<jsp:useBean>语句

    C. bean 文件放在任何目录下都可以被引用

    D. 以上均不对

7. 写一个 bean 时,与布尔逻辑类型的成员变量 XXX 对应的方法是(　　)。

    A. getXXX()   B. setXXX()   C. XXX()   D. isXXX()

8. 在 J2EE 中,test.jsp 文件中有如下一行代码:

```
<jsp:useBean id="user" scope="____" type="com.UserBean" />
```

要使 user 对象在用户对其发出请求时存在,下画线中应填入(　　)。

    A. page   B. request   C. session   D. application

9. 在 JSP 中,使用<jsp:useBean>动作可以将 javaBean 引入 JSP 页面,对 JavaBean 的访问范围不能是(　　)。

    A. page   B. request   C. response   D. application

10. 下面语句与<jsp:getProperty name="aBean" property="jsp"/>等价的是(　　)。

    A. <%=jsp()%>

    B. <%out.print(aBean.getJsp());%>

    C. <%=aBean.setJsp()%>

    D. <%aBean.setJsp();%>

11. 在 JSP 中,以下是有关 jsp:setProperty 和 jsp:getProperty 标记的描述,正确的是(　　)。

    A. <jsp:setProperty>和<jsp:getProperty>标记都必须在<jsp:useBean>的开始标记和结束标记之间

    B. 这两个标记的 name 属性的值必须和<jsp:useBean>动作标记的 id 属性的值相对应

    C. 这两个标记的 name 属性的值可以和<jsp:useBean>动作标记的 id 属性的值不同

    D. 以上均不对

12. test.jsp 文件如下:

```
<body>
<jsp:useBean id ="buffer" scope="page" class="java.lang.StringBuffer"/>
<%buffer.append("ABC"); %>
    buffer is <%=buffer%>
</body>
```

试图运行 test.jsp 时,将发生(　　)。

A. 编译期间发生错误

B. 运行期间抛出异常

C. 运行后,浏览器上显示:buffer is null

D. 运行后,浏览器上显示:buffer is ABC

13. 在 JSP 中使用<jsp:getProperty>标记时,不会出现的属性是(　　)。

　　A. name　　　　B. property　　　　C. value　　　　D. 以上皆不会出现

第 5 章 JSP 访问数据库

本章主要内容

- 使用 JDBC-ODBC 桥接器连接数据库
- 使用纯 Java 数据库驱动程序连接数据库
- Statement、ResultSet 的使用
- 游动查询
- 访问 Excel 电子表格
- 使用连接池
- 其他典型数据库的连接
- 预处理语句的使用

数据库在现在的 Web 应用中扮演着越来越重要的角色。如果没有数据库,很多重要的应用,像电子商务、搜索引擎等都不可能实现。这一章将主要介绍在 JSP 中如何访问关系数据库,如 Oracle、SQL Server、MySQL 和 Microsoft Access 等数据库。

在本章中,新建一个 Web 工程 ch5,本章例子中涉及的 Java 源文件保存在 ch5 的 src 目录中,涉及的 JSP 页面保存在 ch5 的 WebContent 目录中。

5.1 使用 JDBC-ODBC 桥接器连接数据库

JSP 页面中访问数据库,首先要与数据库进行连接,通过连接向数据库发送指令,并获得返回的结果。JDBC 连接数据库有两种常用方式:建立 JDBC-ODBC 桥接器和加载纯 Java 数据库驱动程序。

5.1.1 核心知识

JDBC(Java DataBase Connectivity)是用于运行 SQL 的解决方案,是 Java 运行平台核心类库中的一部分,它由一组标准接口与类组成。我们经常使用 JDBC 完成 3 件事:与指定的数据库建立连接;向已连接的数据库发送 SQL 命令;处理 SQL 命令返回的结果。

ODBC(Open DataBase Connectivity)是由 Microsoft 主导的数据库连接标准,提供了通用的数据库访问平台。但是,使用 ODBC 连接数据库的应用程序移植性较差,因为应用程

序所在的计算机必须提供 ODBC。

使用 JDBC-ODBC 桥接器连接数据库的机制是：将连接数据库的相关信息提供给 JDBC-ODBC 驱动程序，然后转换成 JDBC 接口，以供应用程序使用，而与数据库的连接是由 ODBC 完成的。使用 JDBC-ODBC 桥接器连接数据库如图 5.1 所示。

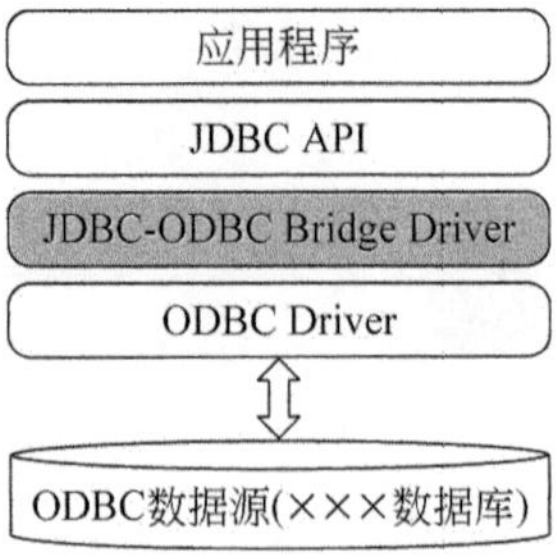

使用 JDBC-ODBC 桥接器连接数据库有如下 3 个步骤。

(1) 建立 JDBC-ODBC 桥接器。
(2) 创建 ODBC 数据源。
(3) 与 ODBC 数据源指定的数据库建立连接。

图 5.1 JDBC-ODBC 桥方式

5.1.2 能力目标

掌握 JDBC-ODBC 桥接器连接数据库的方法。

5.1.3 任务驱动

1. 任务的主要内容

- 创建待连接的 Microsoft Access 数据库；
- 建立 JDBC-ODBC 桥接器；
- 创建 ODBC 数据源；
- 与 ODBC 数据源指定的数据库建立连接；
- 在 JSP 页面中使用 JDBC-ODBC 桥接器连接数据库。

2. 任务模板

请按下列步骤进行操作。

1) 创建待连接的 Microsoft Access 数据库

使用 Microsoft Access 2007 设计一个数据库 goods.accdb，该库中有一张表，表的名字为 goodsInfo，表的说明如图 5.2 所示。

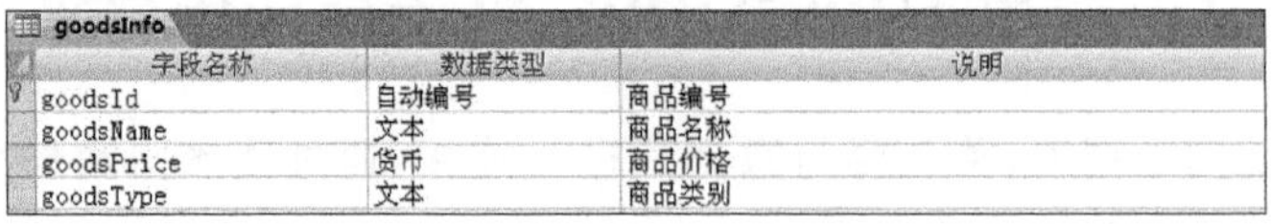

字段名称	数据类型	说明
goodsId	自动编号	商品编号
goodsName	文本	商品名称
goodsPrice	货币	商品价格
goodsType	文本	商品类别

图 5.2 goodsInfo 表的说明

表中的数据如图 5.3 所示。

2) 建立 JDBC-ODBC 桥接器

JDBC 通过 java.lang.Class 类的静态方法 forName 加载 sun.jdbc.odbc.JdbcOdbcDriver 类建立 JDBC-ODBC 桥接器。建立桥接器时可能发生 ClassNotFoundException 异常，必须捕获该异常，建立桥接器的具体代码如下：

```
try{
        Class.forName("sun.jdbc.odbc.JdbcOdbcDriver");
}catch(ClassNotFoundException e){
```

```
        e.printStackTrace();
}
```

3) 创建 ODBC 数据源

创建 ODBC 数据源时，必须保证计算机有 ODBC 系统，Windows 操作系统一般都带有 ODBC 系统。

(1) 打开 ODBC 数据源管理器

打开"控制面板"中的"管理工具"（有些 Windows XP 系统，必须打开"控制面板"中的"性能和维护"选项，才能打开"管理工具"），找到"数据源（ODBC）"图标，双击打开，出现如图 5.4 所示的界面。

图 5.3　goodsInfo 表中的数据　　　　图 5.4　打开 ODBC 数据源管理器

(2) 为数据源选择驱动程序

在图 5.4 所示的界面上选择"用户 DSN"选项卡，单击"添加"按钮，出现为新增的数据源选择驱动程序界面，如图 5.5 所示。因为我们要连接 Microsoft Access 2007 数据库，在此选择 Microsoft Access Driver(*.mdb,*.accdb)选项，然后单击"完成"按钮。

图 5.5　为新增的数据源选择驱动程序

(3) 为数据源起名并找到对应的数据库

单击"完成"按钮后出现设置数据源具体信息的对话框，如图 5.6 所示。在"数据源名"文本框中为数据源起个名字：myGod。在图 5.6 所示界面里单击"选择"按钮为 myGod 数据源选择数据库。

(4) 设置登录名和密码

在图 5.6 所示界面单击"高级"按钮出现设置登录名与密码界面，如图 5.7 所示。这里的用户名和密码都是 firstDB。在图 5.7 所示界面单击"确定"按钮后，再单击图 5.6 所示界面的"确定"按钮就创建了一个新的数据源：myGod。

图 5.6 设置数据源的名字和选择数据库

图 5.7 设置登录名和密码

4) 与 ODBC 数据源指定的数据库建立连接

首先,使用 java.sql 包中的 Connection 类声明一个连接对象 con,然后使用 java.sql 包中的 DriverManager 类调用静态方法 getConnection 创建连接对象 con,代码如下:

```
Connection con=DriverManager.getConnection("jdbc:odbc:数据源名字","登录名","密码");
```

如果没有给数据源设置登录名和密码,那么连接形式是:

```
Connection con=DriverManager.getConnection("jdbc:odbc:数据源名字","","");
```

建立连接时应捕获 SQLException 异常,例如与数据源 myGod 指定的数据库 goods.accdb 建立连接,代码如下:

```
try{
     Connection con = DriverManager.getConnection("jdbc:odbc:myGod","firstDB","firstDB");
}catch(SQLException e){
     e.printStackTrace();
}
```

5) 在 JSP 页面中使用 JDBC-ODBC 桥接器连接数据库

编写一个 JSP 页面 example5_1.jsp,该页面中的 Java 程序片代码使用 JDBC-ODBC 桥接器连接到数据源 myGod,查询 goodsInfo 表中的全部记录。页面运行效果如图 5.8 所示。

图 5.8 使用 JDBC-ODBC 桥接器连接数据库

example5_1.jsp

```
<%@page language="java" contentType="text/html; charset=GBK" pageEncoding="GBK"%>
<%@page import="java.sql.*"%>
<html>
<head>
<title>example5_1.jsp</title>
</head>
<body bgcolor="lightgreen">
    <%
```

```
            Connection con=null;
            Statement st=null;
            ResultSet rs=null;
            try {
                Class.forName("sun.jdbc.odbc.JdbcOdbcDriver");
            } catch (ClassNotFoundException e) {
                e.printStackTrace();
            }
            try {
                con= DriverManager. getConnection ( " jdbc: odbc: myGod "," firstDB ",
            "firstDB");
                st=con.createStatement();
                rs=st.executeQuery("select *  from goodsInfo");
                out.print("<table border=1>");
                out.print("<tr>");
                    out.print("<th>商品编号</th>");
                    out.print("<th>商品名称</th>");
                    out.print("<th>商品价格</th>");
                    out.print("<th>商品类别</th>");
                out.print("</tr>");
                while(rs.next()){
                    out.print("<tr>");
                        out.print("<td>"+rs.getString(1)+"</td>");
                        out.print("<td>"+rs.getString(2)+"</td>");
                        out.print("<td>"+rs.getString(3)+"</td>");
                        out.print("<td>"+rs.getString(4)+"</td>");
                    out.print("</tr>");
                }
                out.print("</table>");
            } catch (SQLException e) {
                e.printStackTrace();
            } finally{
                try{
                    if(rs!=null){
                        rs.close();
                    }
                    if(st!=null){
                        st.close();
                    }
                    if(con!=null){
                        con.close();
                    }
                }catch (SQLException e) {
                    e.printStackTrace();
                }
            }
      %>
</body>
</html>
```

5.1.4 实践环节

（1）参考本节任务中的主要内容，创建数据源 mySky，该数据源指定的数据库是 goods.accdb。

（2）编写一个 JSP 页面 practice5_1.jsp，该页面中的 Java 程序片代码使用 JDBC-ODBC 桥接器连接到数据源 mySky，查询 goodsInfo 表中 goodsPrice 字段值大于 100 的全部记录。页面运行效果如图 5.9 所示。

图 5.9 practice5_1.jsp 页面运行效果

5.2 使用纯 Java 数据库驱动程序连接数据库

5.2.1 核心知识

应用程序除了使用 JDBC-ODBC 桥接器连接数据库外，通常还使用 JDBC API 调用本地的纯 Java 数据库驱动程序和相应的数据库建立连接，如图 5.10 所示。

使用纯 Java 数据库驱动程序连接数据库，需要经过如下两个步骤。

（1）注册纯 Java 数据库驱动程序。

（2）与指定的数据库建立连接。

下面以 Oracle 10g 为例，讲解如何使用纯 Java 数据库驱动程序连接数据库。

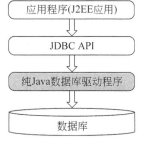

图 5.10 使用纯 Java 数据库驱动程序

1. 注册纯 Java 数据库驱动程序

每种数据库都配有自己的纯 Java 数据库驱动程序。Oracle 10g 的纯 Java 驱动程序一般位于数据库安装目录"\oracle\product\10.2.0\db_1\jdbc\lib"下，名为 classes12.jar。

为了连接 Oracle 10g 数据库，可以将 classes12.jar 文件复制到 Web 应用程序的/WEB-INF/lib 文件夹中。然后，通过 java.lang.Class 类的 forName()，动态注册 Oracle 10g 的纯 Java 数据库驱动程序，代码如下：

```
try {
    Class.forName("oracle.jdbc.driver.OracleDriver");
} catch (ClassNotFoundException e) {
    e.printStackTrace();
}
```

2. 与指定的数据库建立连接

与 Oracle 数据库建立连接的代码如下：

```
try {
    Connection con=DriverManager.getConnection("jdbc:oracle:thin:@主机:端口号:数据库名","用户名","密码");
```

```
    } catch (SQLException e) {
        e.printStackTrace();
    }
```

其中，主机是安装 Oracle 服务器的 IP 地址，如果是本机则为 localhost；Oracle 默认端口号为 1521；Oracle 默认数据库名为 orcl；用户名和密码是访问 Oracle 服务器的用户权限。

本章中后续的例子(除 5.5 节与 5.7 节)均采用纯 Java 数据库驱动程序连接 Oracle 10g。

5.2.2 能力目标

掌握纯 Java 数据库驱动程序连接数据库的方法。

5.2.3 任务驱动

1. 任务的主要内容

编写一个 JSP 页面 example5_2.jsp，该页面的 Java 程序片代码使用纯 Java 驱动程序连接 Oracle 数据库，查询 goodsInfo 表中的全部记录。创建 goodsInfo 表的 SQL 语句如下：

```
create table goodsinfo (
    goodsId number(4) not null,
    goodsName varchar(50) not null,
    goodsPrice number(7,2) not null,
    goodsType varchar(10) not null,
    constraint pk_goodsinfo primary key (goodsId)
);
insert into goodsinfo values(1,'牙膏',12,'日用品');
insert into goodsinfo values(2,'冰箱',2500,'电器');
insert into goodsinfo values(3,'蛋糕',28,'食品');
insert into goodsinfo values(4,'苹果',48,'水果');
insert into goodsinfo values(5,'上衣',1800,'服装');
insert into goodsinfo values(6,'书包',78,'文具');
commit;
```

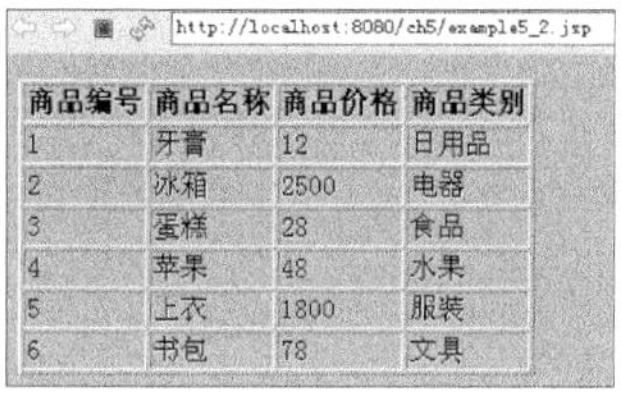

商品编号	商品名称	商品价格	商品类别
1	牙膏	12	日用品
2	冰箱	2500	电器
3	蛋糕	28	食品
4	苹果	48	水果
5	上衣	1800	服装
6	书包	78	文具

图 5.11　使用纯 Java 驱动程序连接 Oracle 数据库

页面运行效果如图 5.11 所示。

2. 任务的代码模板

将下列 example5_2.jsp 中的【代码】替换为真正的 JSP 的代码。

example5_2.jsp

```
<%@page language="java" contentType="text/html; charset=GBK" pageEncoding=
"GBK"%>
<%@page import="java.sql.*"%>
<html>
<head>
<title>example5_2.jsp</title>
</head>
<body bgcolor="LightPink">
    <%
        Connection con=null;
```

```
            Statement st=null;
            ResultSet rs=null;
            try{
                【代码】                           //注册 Oracle 的纯 Java 驱动程序
            } catch (ClassNotFoundException e) {
                e.printStackTrace();
            }
            try{
                con= DriverManager.getConnection("jdbc:oracle:thin:@localhost:
                1521:orcl",
                    "system","system");
                st=con.createStatement();
                rs=st.executeQuery("select * from goodsInfo");
                out.print("<table border=1>");
                out.print("<tr>");
                    out.print("<th>商品编号</th>");
                    out.print("<th>商品名称</th>");
                    out.print("<th>商品价格</th>");
                    out.print("<th>商品类别</th>");
                out.print("</tr>");
                while(rs.next()){
                    out.print("<tr>");
                        out.print("<td>"+rs.getString(1)+"</td>");
                        out.print("<td>"+rs.getString(2)+"</td>");
                        out.print("<td>"+rs.getString(3)+"</td>");
                        out.print("<td>"+rs.getString(4)+"</td>");
                    out.print("</tr>");
                }
                out.print("</table>");
            } catch (SQLException e) {
                e.printStackTrace();
            } finally{
                try{
                    if(rs!=null){
                        rs.close();
                    }
                    if(st!=null){
                        st.close();
                    }
                    if(con!=null){
                        con.close();
                    }
                }catch (SQLException e) {
                    e.printStackTrace();
                }
            }
    %>
</body>
</html>
```

3. 任务小结或知识扩展

应用程序连接 Oracle 数据库时，必须事先启动 Oracle 服务器的 OracleServiceORCL 和 OracleOraDb10g_home1TNSListener 两个服务，否则会抛出连接异常。

从任务中可以看出编写程序访问数据库需要有以下几个步骤。

（1）导入 java.sql 包

所有与数据库有关的对象和方法都在 java.sql 包中，包 java.sql 包含了用 Java 操作关系数据库的类和接口。因此在使用 JDBC 的程序中必须加入 import java.sql.*。

（2）加载驱动程序

在该任务中使用了 Class 类（java.lang 包）中的方法 forName，来装入该驱动程序的类定义 oracle.jdbc.driver.OracleDriver，从而创建了该驱动程序的一个实例。

（3）连接数据库

完成上述操作后，就可以连接一个特定的数据库了。这需要创建 Connection 类的一个实例，并使用 DriverManager 的方法 getConnection 尝试与指定的数据库建立连接。代码如下：

```
con=DriverManager.getConnection("jdbc:oracle:thin:@localhost:1521:orcl",
"system","system");
```

（4）访问数据库

访问数据库，需要先用 Connection 类的 createStatement 方法从指定的数据库连接得到一个 Statement 的实例，然后用这个实例的相应方法来执行一条 SQL 语句。代码如下：

```
st=con.createStatement();
rs=st.executeQuery("select * from goodsInfo");
```

（5）处理返回的结果集

ResultSet 对象是 JDBC 中比较重要的一个对象，几乎所有的查询操作都将数据作为 ResultSet 对象返回。处理结果集 ResultSet 对象的代码如下：

```
while(rs.next()){
    out.print("<tr>");
    out.print("<td>"+rs.getString(1)+"</td>");
    out.print("<td>"+rs.getString(2)+"</td>");
    out.print("<td>"+rs.getString(3)+"</td>");
    out.print("<td>"+rs.getString(4)+"</td>");
    out.print("</tr>");
}
```

（6）关闭数据库连接，释放资源

对数据库的操作完成之后，要及时关闭 ResultSet 对象、Statement 对象和数据库连接对象 Connection，从而释放占用的资源，这就要用到 close 方法。代码如下：

```
rs.close();
st.close();
con.close();
```

关闭的顺序从先到后依次为 ResultSet 对象、Statement 对象和 Connection 对象。

4. 代码模板的参考答案

【代码】：`Class.forName("oracle.jdbc.driver.OracleDriver");`

5.2.4 实践环节

编写一个 JSP 页面 practice5_2.jsp，该页面中的 Java 程序片代码使用纯 Java 数据库驱动程序连接 Oracle 数据库，查询 goodsInfo 表中 goodsPrice 字段值大于 10 并小于 50 的全部记录。页面运行效果如图 5.12 所示。

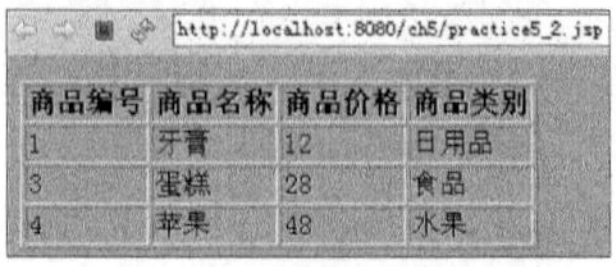

图 5.12 practice5_2.jsp 页面运行效果

5.3 Statement、ResultSet 的使用

5.3.1 核心知识

与数据库建立连接之后，接下来若要执行 SQL 语句，需要有以下几个步骤。

1. 创建 Statement 对象

Statement 对象代表一条发送到数据库执行的 SQL 语句。由已创建的 Connection 对象 con 调用 createStatement() 方法来创建 Statement 对象，代码如下：

```
Statement smt=con.createStatement();
```

2. 执行 SQL 语句

创建 Statement 对象之后，可以使用 Statement 对象调用 executeUpdate(String sql)、executeQuery(String sql) 等方法来执行 SQL 语句。

executeUpdate(String sql) 方法主要用于执行 INSERT、UPDATE 或 DELETE 语句以及 SQL DDL 语句，例如 CREATE TABLE 和 DROP TABLE。该方法返回一个整数（代表被更新的行数），对于 CREATE TABLE 和 DROP TABLE 等不操作行的指令，返回零。

executeQuery(String sql) 方法则是用于执行 SELECT 等查询数据库的 SQL 语句，该方法返回 ResultSet 对象，代表查询的结果。

3. 处理返回的 ResultSet 对象

ResultSet 对象是 executeQuery(String sql) 方法的返回值，被称为结果集，它代表符合 SQL 语句条件的所有行。ResultSet 对象调用 next() 方法移动到下一个数据行（顺序查询），当数据行存在时，next() 方法返回 true，否则返回 false。获得一行数据后，ResultSet 对象可以使用 getXXX() 方法获得字段值，getXXX() 方法都提供依字段名称取得的数据，或是依字段顺序取得数据的方法。

5.3.2 能力目标

能够灵活使用 Statement 与 ResultSet 对象对数据库进行增删改查。

5.3.3 任务驱动

1. 任务的主要内容

编写两个 JSP 页面：addGoods.jsp 和 showAllGoods.jsp。用户可以在 addGoods.jsp 页面中输入信息后，单击"添加"按钮把信息添加到 goodsInfo 表中。然后，在 showAllGoods.jsp 页面中显示所有商品信息。在该任务中需要编写一个 bean(GoodsBean.java)，用来实现添加和查询记录。页面运行效果如图 5.13(a)、图 5.13(b)所示。

(a) 添加记录　　　　　　　　　　　(b) 查询记录

图 5.13　5.3 节任务的页面运行效果图

2. 任务的代码模板

将下列 GoodsBean.java 中的【代码】替换为真正的 Java 的代码。

addGoods.jsp

```
<%@page language="java" contentType="text/html; charset=GBK" pageEncoding=
"GBK"%>
<html>
<head>
<title>addGoods.jsp</title>
</head>
<body>
    <h4>商品编号是主键,不能重复,每个信息都必须输入!</h4>
    <form action="showAllGoods.jsp" method="post">
    <table border="1">
        <tr>
            <td>商品编号:</td>
            <td><input type="text" name="goodsId"/></td>
        </tr>

        <tr>
            <td>商品名称:</td>
            <td><input type="text" name="goodsName"/></td>
        </tr>

        <tr>
            <td>商品价格:</td>
            <td><input type="text" name="goodsPrice"/></td>
        </tr>

        <tr>
```

```
            <td>商品类型:</td>
            <td>
                <select name="goodsType">
                    <option value="日用品">日用品
                    <option value="电器">电器
                    <option value="食品">食品
                    <option value="水果">水果
                    <option value="服装">服装
                    <option value="文具">文具
                    <option value="其他">其他
                </select>
            </td>
        </tr>

        <tr>
            <td><input type="submit" value="添加"></td>
            <td><input type="reset" value="重置"></td>
        </tr>
    </table>
    </form>
</body>
</html>
```

showAllGoods.jsp

```
<%@page language="java" contentType="text/html; charset=GBK" pageEncoding="GBK"%>
<%@page import="bean.GoodsBean" %>
<html>
<head>
<title>showAllGoods.jsp</title>
</head>
<body>
    <%
        request.setCharacterEncoding("GBK");
    %>
    <jsp:useBean id="goods" class="bean.GoodsBean" scope="page"></jsp:useBean>
    <jsp:setProperty property="*" name="goods"/>
    <%
        goods.addGoods();                              //添加商品
    %>
    <jsp:getProperty property="queryResult" name="goods"/><!--获得查询结果 -->
</body>
</html>
```

GoodsBean.java

```
package bean;
import java.sql.*;
public class GoodsBean {
    int goodsId;
```

```java
String goodsName;
double goodsPrice;
String goodsType;
StringBuffer queryResult;                    //查询结果
public GoodsBean(){

}
public int getGoodsId() {
    return goodsId;
}
public void setGoodsId(int goodsId) {
    this.goodsId=goodsId;
}
public String getGoodsName() {
    return goodsName;
}
public void setGoodsName(String goodsName) {
    this.goodsName=goodsName;
}
public double getGoodsPrice() {
    return goodsPrice;
}
public void setGoodsPrice(double goodsPrice) {
    this.goodsPrice=goodsPrice;
}
public String getGoodsType() {
    return goodsType;
}
public void setGoodsType(String goodsType) {
    this.goodsType=goodsType;
}
//添加记录
public void addGoods(){
    Connection con=null;
    Statement st=null;
    try {
        Class.forName("oracle.jdbc.driver.OracleDriver");
    } catch (ClassNotFoundException e) {
        e.printStackTrace();
    }
    try {
        con= DriverManager.getConnection ("jdbc:oracle:thin:@localhost:1521:orcl",
                "system","system");
        【代码1】                        //创建 Statement 对象
        String addSql="insert into goodsInfo
            values ("+ goodsId +", '" + goodsName +"','" + goodsPrice +", '" +
            goodsType+"')";
        【代码2】                        //执行 insert 语句
    }catch (SQLException e) {
        e.printStackTrace();
```

```
        }finally{
            try{
                if(st!=null){
                    st.close();
                }
                if(con!=null){
                    con.close();
                }
            }catch (SQLException e) {
                e.printStackTrace();
            }
        }
    }
    //查询记录
    public StringBuffer getQueryResult(){
        queryResult=new StringBuffer();
        Connection con=null;
        Statement st=null;
        ResultSet rs=null;
        try {
            Class.forName("oracle.jdbc.driver.OracleDriver");
        } catch (ClassNotFoundException e) {
            e.printStackTrace();
        }
        try {
             con = DriverManager.getConnection("jdbc:oracle:thin:@localhost:
             1521:orcl",
                    "system","system");
            【代码 3】                    //创建 Statement 对象
            String selectSql="select * from goodsInfo";
            【代码 4】                    //执行 select 语句
            queryResult.append("<table border=1>");
                queryResult.append("<tr>");
                    queryResult.append("<th>goodsId</th>");
                    queryResult.append("<th>goodsName</th>");
                    queryResult.append("<th>goodsPrice</th>");
                    queryResult.append("<th>goodsType</th>");
                queryResult.append("</tr>");
            while(rs.next()){
                queryResult.append("<tr>");
                    queryResult.append("<td>"+rs.getString(1)+"</td>");
                    queryResult.append("<td>"+rs.getString(2)+"</td>");
                    queryResult.append("<td>"+rs.getString(3)+"</td>");
                    queryResult.append("<td>"+rs.getString(4)+"</td>");
                queryResult.append("</tr>");
            }
            queryResult.append("</table>");
        }catch (SQLException e) {
            e.printStackTrace();
        }finally{
            try{
                if(rs!=null){
```

```
                rs.close();
            }
            if(st!=null){
                st.close();
            }
            if(con!=null){
                con.close();
            }
        }catch (SQLException e) {
            e.printStackTrace();
        }
    }
    return queryResult;
  }
}
```

3. 任务小结或知识扩展

ResultSet 对象自动维护指向其当前数据行的游标。每调用一次 next()方法,游标就向下移动一行。最初它位于结果集的第一行之前,因此第一次调用 next()方法,将把游标置于第一行上,使它成为当前行。随着每次调用 next()方法,游标依次向下移动一行,按照从上至下的顺序获取 ResultSet 行,实现顺序查询。

ResultSet 对象包含 SQL 语句的执行结果。它通过一套 get()方法对这些行中的数据进行访问,即使用 getXXX()方法获得数据。get()方法很多,究竟用哪一个 getXXX()方法,由列的数据类型来决定。使用 getXXX()方法时,需要注意以下两点。

(1) 无论列是何种数据类型,都可以使用 getString(int columnIndex)或 getString (String columnName)方法获得列值的字符串表示。

(2) 在使用 getString(int columnIndex)方法查看一行记录时,不允许颠倒顺序,例如不允许:

```
rs.getString(2);
rs.getString(1);
```

4. 代码模板的参考答案

【代码 1】:st=con.createStatement();
【代码 2】:st.executeUpdate(addSql);
【代码 3】:st=con.createStatement();
【代码 4】:rs=st.executeQuery(selectSql);

5.3.4 实践环节

编写两个 JSP 页面:inputQuery.jsp 和 showGoods.jsp。用户可以在 inputQuery.jsp 页面输入查询条件后,单击"查询"按钮。然后,在 showGoods.jsp 页面中显示符合查询条件的商品信息。在本节任务的 bean(GoodsBean.java)中添加一个方法 getQueryResultBy()实现该题的条件查询功能。页面运行效果如图 5.14(a)、图 5.14(b)所示。

(a) 输入条件 　　　　(b) 符合查询条件的记录

图 5.14　5.3 节实践环节页面运行效果图

5.4　游　动　查　询

5.4.1　核心知识

有时候需要结果集的游标前后移动,这时可使用滚动结果集。为了获得滚动结果集,必须先用下面的方法得到一个 Statement 对象。

```
Statement st=con.createStatement(int type, int concurrency);
```

根据 type 和 concurrency 的取值,当执行 ResultSet rs＝st.executeQuery(String sql)时,会返回不同类型的结果集。

type 的取值决定滚动方式,它可以取：

- ResultSet.TYPE_FORWORD_ONLY　表示结果集只能向下滚动。
- ResultSet.TYPE_SCROLL_INSENSITIVE　表示结果集可以上下滚动,当数据库变化时,结果集不变。
- ResultSet.TYPE_SCROLL_SENSITIVE　表示结果集可以上下滚动,当数据库变化时,结果集同步改变。

concurrency 取值表示是否可以用结果集更新数据库,它的取值是：

- ResultSet.CONCUR_READ_ONLY　表示不能用结果集更新数据库的表。
- ResultSet.CONCUR_UPDATETABLE　表示能用结果集更新数据库的表。

5.4.2　能力目标

能够灵活使用滚动结果集进行游动查询。

5.4.3　任务驱动

1. 任务的主要内容

编写一个 JSP 页面 randomQuery.jsp,查询 goodsInfo 表中的全部记录,并将结果逆序输出,最后单独输出第 4 条记录。运行结果如图 5.15 所示。

2. 任务的代码模板

将下列 randomQuery.jsp 中的【代码】替换为真正的 JSP 的代码。

图 5.15　随机查询记录

randomQuery.jsp

```jsp
<%@page language="java" contentType="text/html; charset=GBK" pageEncoding="GBK"%>
<%@page import="java.sql.*"%>
<html>
<head>
<title>randomQuery.jsp</title>
</head>
<body>
    <%
        Connection con=null;
        Statement st=null;
        ResultSet rs=null;
        try {
            Class.forName("oracle.jdbc.driver.OracleDriver");
        } catch (ClassNotFoundException e) {
            e.printStackTrace();
        }
        try {
            con= DriverManager.getConnection("jdbc:oracle:thin:@localhost:1521:orcl", "system",
                    "system");
            //创建st对象,该对象获得的结果集和数据库同步变化,但不能用结果集更新表
            【代码】
            //返回可滚动的结果集
            rs=st.executeQuery("SELECT * FROM goodsInfo");
            //将游标移动到最后一行
            rs.last();
            //获取最后一行的行号
            int lownumber=rs.getRow();
            out.print("该表共有" +lownumber +"条记录");
            out.print("<BR>现在逆序输出记录:");
            out.print("<Table Border=1>");
            out.print("<TR>");
            out.print("<TH>商品编号</TH>");
            out.print("<TH>商品名称</TH>");
            out.print("<TH>商品价格</TH>");
            out.print("<TH>商品类别</TH>");
            out.print("</TR>");
            //为了逆序输出记录,需将游标移动到最后一行之后
            rs.afterLast();
            while (rs.previous()) {
                out.print("<TR>");
                out.print("<TD >" +rs.getString(1) +"</TD>");
                out.print("<TD >" +rs.getString(2) +"</TD>");
                out.print("<TD >" +rs.getString(3) +"</TD>");
                out.print("<TD >" +rs.getString(4) +"</TD>");
                out.print("</TR>");
            }
```

```
                out.print("</Table>");
                out.print("<BR>单独输出第 4 条记录<BR>");
                rs.absolute(4);
                out.print(rs.getString(1) +" ");
                out.print(rs.getString(2) +" ");
                out.print(rs.getString(3) +" ");
                out.print(rs.getString(4));
            } catch (SQLException e) {
                e.printStackTrace();
            } finally {
                try {
                    if (rs !=null) {
                        rs.close();
                    }
                    if (st !=null) {
                        st.close();
                    }
                    if (con !=null) {
                        con.close();
                    }
                } catch (SQLException e) {
                    e.printStackTrace();
                }
            }
        %>
    </body>
</html>
```

3. 任务小结或知识扩展

游动查询经常用到 ResultSet 类的下述方法。

- public boolean absolute(int row)：将游标移到参数 row 指定的行。如果 row 取负值，就是倒数的行，如-1 表示最后一行。当移到最后一行之后或第一行之前时，该方法返回 false。
- public void afterLast()：将游标移到结果集的最后一行之后。
- public void beforeFirst()：将游标移到结果集的第一行之前。
- public void first()：将游标移到结果集的第一行。
- public int getRow()：得到当前游标所指定的行号，如果没有行，则返回 0。
- public boolean isAfterLast()：判断游标是不是在结果集的最后一行之后。
- public boolean isBeforeFirst()：判断游标是不是在结果集的第一行之前。
- public void last()：将游标移到结果集的最后一行。
- public boolean previous()：将游标向上移动(和 next()方法相反)，当移到结果集的第一行之前时返回 false。

4. 代码模板的参考答案

【代码】：st=con.createStatement(ResultSet.TYPE_SCROLL_SENSITIVE,

```
ResultSet.CONCUR_READ_ONLY);
```

5.4.4 实践环节

编写一个 JSP 页面 practice5_4.jsp,查询 goodsInfo 表中的记录,并逆序输出偶数行的记录。运行结果如图 5.16 所示。

图 5.16 逆序输出偶数行的记录

5.5 访问 Excel 电子表格

5.5.1 核心知识

当需要对 Excel 电子表格的内容进行增加、修改、查询和删除时,可以使用 JDBC-ODBC 桥接器的方式访问 Excel 电子表格。但访问 Excel 电子表格和访问关系数据库有所不同。访问 Excel 电子表格要经过以下两个步骤。

1. 创建 Excel 电子表格

使用 Excel 2007 创建一个名为 student.xlsx 的电子表格,该电子表格中有个名为 studentScore 的工作表。student.xlsx 电子表格如图 5.17 所示。

2. 创建数据源

创建 Excel 电子表格之后,就可以为它创建数据源了。数据源的名字是 student,为数据源选择的驱动程序是 Microsoft Excel Driver。如果想通过 JDBC-ODBC 修改 Excel 电子表格,在设置数据源时,单击"选项"按钮,将"只读"属性设置为未选中状态,如图 5.18 所示。

图 5.17 student.xlsx 电子表格

图 5.18 创建数据源

注:访问电子表格时,把其中的工作表看成是数据库中的表,如 student.xlsx 中的 studentScore 工作表。使用 SQL 语句对工作表中的数据进行增加、删除、修改和查询时,要在表名前加"[",在表名后加"$]",如 select * from [studentScore$]。

5.5.2 能力目标

能够灵活使用 JDBC-ODBC 桥接器的方式访问 Excel 电子表格。

5.5.3 任务驱动

1. 任务的主要内容

编写一个 JSP 页面 readExcel.jsp，在该页面的 Java 程序片中首先增加一条记录到 studentScore 工作表中，然后修改某条记录，最后再查询全部记录。页面运行效果如图 5.19 所示。

2. 任务的代码模板

将下列 readExcel.jsp 中的【代码】替换为真正的 JSP 的代码。

图 5.19 增加、修改和查询 Excel 表的记录

readExcel.jsp

```jsp
<%@ page language="java" contentType="text/html; charset=GBK" pageEncoding="GBK"%>
<%@ page import="java.sql.*"%>
<html>
<head>
<title>readExcel.jsp</title>
</head>
<body bgcolor="lightgreen">
    <%
        Connection con=null;
        Statement st=null;
        ResultSet rs=null;
        try {
            Class.forName("sun.jdbc.odbc.JdbcOdbcDriver");
        } catch (ClassNotFoundException e) {
            e.printStackTrace();
        }
        try {
            con=DriverManager.getConnection("jdbc:odbc:student","", "");
            st=con.createStatement();
            //添加记录
            st.executeUpdate("insert into【代码1】values('201212009','sss','nv',73)");
            //修改记录
            st.executeUpdate("update【代码2】set 性别='女' where 学号='201288803'");
            //查询全部记录
            rs=st.executeQuery("select * from【代码3】");
            out.print("<table border=1>");
            out.print("<tr>");
                out.print("<th>学号</th>");
                out.print("<th>姓名</th>");
                out.print("<th>性别</th>");
                out.print("<th>JSP 成绩</th>");
            out.print("</tr>");
            while(rs.next()){
                out.print("<tr>");
                    out.print("<td>"+rs.getString(1)+"</td>");
                    out.print("<td>"+rs.getString(2)+"</td>");
```

```
                out.print("<td>"+rs.getString(3)+"</td>");
                out.print("<td>"+rs.getString(4)+"</td>");
            out.print("</tr>");
        }
        out.print("</table>");
    } catch (SQLException e) {
        e.printStackTrace();
    }finally{
        try{
            if(rs!=null){
                rs.close();
            }
            if(st!=null){
                st.close();
            }
            if(con!=null){
                con.close();
            }
        }catch (SQLException e) {
            e.printStackTrace();
        }
    }
%>
</body>
</html>
```

3. 任务小结或知识扩展

一个 Excel 电子表格可以有多张工作表，我们使用 JDBC-ODBC 可以访问该电子表格中的任何一张工作表，就像访问一个数据库中的任意一张表一样。

4. 代码模板的参考答案

【代码 1】：[studentScore$]
【代码 2】：[studentScore$]
【代码 3】：[studentScore$]

5.5.4 实践环节

在 student. xlsx 电子表格中新建一个工作表 empTable(如图 5.20 所示)，编写一个 JSP 页面 practice5_5. jsp，在该页面中显示 empTable 工作表中的所有记录(如图 5.21 所示)。

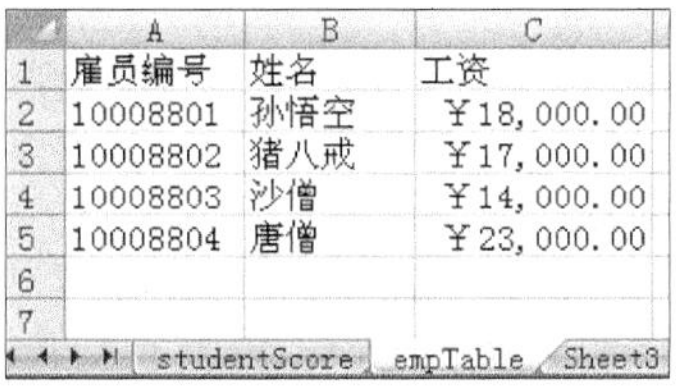

	A	B	C
1	雇员编号	姓名	工资
2	10008801	孙悟空	￥18,000.00
3	10008802	猪八戒	￥17,000.00
4	10008803	沙僧	￥14,000.00
5	10008804	唐僧	￥23,000.00
6			
7			

图 5.20　工作表 empTable

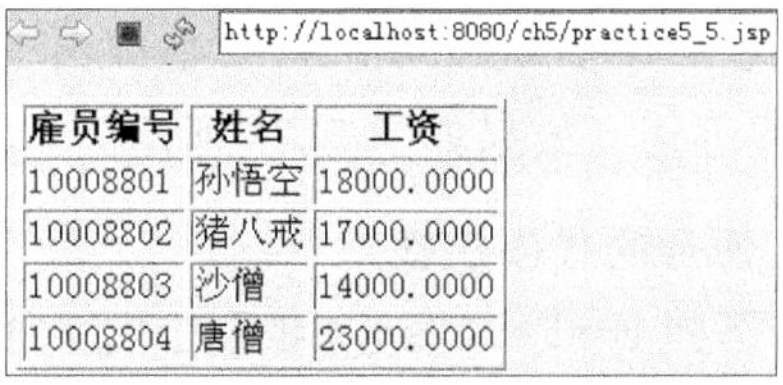

雇员编号	姓名	工资
10008801	孙悟空	18000.0000
10008802	猪八戒	17000.0000
10008803	沙僧	14000.0000
10008804	唐僧	23000.0000

图 5.21　practice5_5.jsp 页面

5.6 使用连接池

5.6.1 核心知识

与数据库建立连接是一个费时的活动,每次都得花费一定的时间,而且系统还要分配内存资源。这个时间对于一次或几次数据库操作,或许感觉不出系统有多大的开销。可是对于大型电子商务网站来说,同时有几百人甚至几千人频繁地进行数据库连接操作势必占用很多的系统资源,网站的响应速度必定将会下降,严重时甚至会造成服务器的崩溃。因此,合理地建立数据库连接是非常重要的。

数据库连接池的基本思想是:为数据库连接建立一个"缓冲池"。预先在"缓冲池"中放入一定数量的连接,当需要建立数据库连接时,只需从"缓冲池"中取出一个,使用完毕之后再放回去。可以通过设定连接池最大连接数来防止系统无限度地与数据库连接。更为重要的是,通过连接池的管理机制监视数据库连接的数量及使用情况,为系统开发、测试和性能调整提供依据。其工作原理如图 5.22 所示。

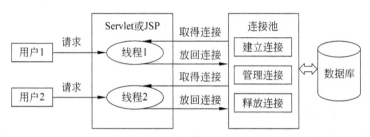

图 5.22 连接池的原理

5.6.2 能力目标

了解连接池的工作原理,灵活使用连接池连接数据库。

5.6.3 任务驱动

1. 任务的主要内容

编写一个 JSP 页面 conPool.jsp,在该页面中使用 scope 设为 application 的 bean(由 ConnectionPool 类负责创建)。在创建该 bean 时,将建立一定数量的连接对象。因此,所有的用户将共享这些连接对象。在 JSP 页面中使用 bean 获得一个连接对象,然后使用该连接对象访问数据库中的 goodsInfo 表(查询出商品价格大于 500 的商品)。页面运行效果如图 5.23 所示。

图 5.23 使用连接池连接数据库

2. 任务的代码模板

将下列 conPool.jsp 中的【代码】替换为真正的 JSP 的代码。

ConnectionPool.java

```
package db.connection.pool;
```

```java
import java.sql.*;
import java.util.ArrayList;
public class ConnectionPool {
    //存放 Connection 对象的数组,数组被看成连接池
    ArrayList<Connection>list=new ArrayList<Connection>();
    //构造方法,创建 15 个连接对象,放到连接池中
    public ConnectionPool(){
        try {
            Class.forName("oracle.jdbc.driver.OracleDriver");
        } catch (ClassNotFoundException e) {
            e.printStackTrace();
        }
        for(int i=0;i<15;i++){
            try {
                Connection  con=DriverManager.getConnection("jdbc:oracle:
                  thin:@localhost:1521:
                orcl","system","system");
                list.add(con);
            } catch (SQLException e) {
                //TODO Auto-generated catch block
                e.printStackTrace();
            }
        }
    }
    //从连接池中取出一个连接对象
    public synchronized Connection getOneCon(){
        if(list.size()>0){
            //删除数组的第一个元素,并返回该元素中的连接对象
            return list.remove(0);
        }else{
            //连接对象被用完
            return null;
        }
    }
    //把连接对象放回连接池中
    public synchronized void releaseCon(Connection con){
        list.add(con);
    }
}
```

conPool.jsp

```
<%@page language="java" contentType="text/html; charset=GBK" pageEncoding="GBK"%>
<%@page import="java.sql.*"%>
<%@page import="db.connection.pool.*"%>
<html>
<head>
<title>使用连接池连接数据库</title>
</head>
<jsp:useBean id="conpool" class="db.connection.pool.ConnectionPool" scope=
```

```jsp
    "application"/>
<body bgcolor="AliceBlue">
<%
    Connection con=null;
    Statement st=null;
    ResultSet rs=null;
    try{
        【代码1】        //使用conpool对象调用getOneCon方法从连接池中获得一个连接对象
        if(con==null){
            out.print("人数过多,稍后访问");
            return;
        }
        st=con.createStatement();
        rs=st.executeQuery("select * from goodsInfo where goodsPrice>500");
        out.print("<table border=1>");
            out.print("<tr>");
                out.print("<th>商品编号</th>");
                out.print("<th>商品名称</th>");
                out.print("<th>商品价格</th>");
                out.print("<th>商品类别</th>");
            out.print("</tr>");
        while(rs.next()){
            out.print("<tr>");
                out.print("<td>"+rs.getString(1)+"</td>");
                out.print("<td>"+rs.getString(2)+"</td>");
                out.print("<td>"+rs.getString(3)+"</td>");
                out.print("<td>"+rs.getString(4)+"</td>");
            out.print("</tr>");
        }
        out.print("</table>");
    }catch(SQLException e){
        e.printStackTrace();
    }finally{
        try{
            if(rs!=null){
                rs.close();
            }
            if(st!=null){
                st.close();
            }
            if(con!=null){
                【代码2】 //使用conpool对象调用releaseCon方法把连接对象放回连接池中
            }
        }catch(SQLException e){
            e.printStackTrace();
        }
    }
```

```
%>
</body>
</html>
```

3. 任务小结或知识扩展

我们在打开一个新的浏览器窗口运行 conPool.jsp 页面时,会发现这一次访问的速度要比第一次快得多,而且也比上几节中访问 JSP 页面的速度要快。这是因为在第一次访问时连接池中没有可用连接,因此页面要等待创建一个新的连接,但是在第二次访问时连接池中已有一个可用的连接了,可以直接使用这个连接来访问数据库。

4. 代码模板的参考答案

【代码 1】:con=conpool.getOneCon();
【代码 2】:conpool.releaseCon(con);

5.6.4 实践环节

编写一个 JSP 页面 pratice5_6.jsp,在该页面中使用和任务中同样的 bean 获得一个数据库连接对象,然后使用该连接对象查询 goodsInfo 表中的全部记录。

5.7 其他典型数据库的连接

5.7.1 核心知识

使用 JDBC-ODBC 桥接器的方式连接不同类型数据库的程序流程是类似的,只是在为 ODBC 数据源选择驱动程序时,选择对应的驱动程序即可,但有的数据库 ODBC 不支持,如 MySQL 数据库。

使用纯 Java 数据库驱动程序连接不同类型数据库的程序流程和框架是基本相同的,需要重点关注的是,在连接各数据库时驱动程序加载部分的代码和连接部分的代码。下面分别介绍直接加载纯 Java 数据库驱动程序连接 SQL Server 2005 和 MySQL 5.5。这里假定要访问的数据库名为 mydatabase。

1. 连接 SQL Server 2005

(1) 获取纯 Java 数据库驱动程序

可以登录微软的官方网站 http://www.microsoft.com 下载 Microsoft SQL Server 2005 JDBC Driver 1.2,解压 Microsoft SQL Server 2005 jdbc driver1.2.exe 后得到 sqljdbc.jar 文件。然后,把 sqljdbc.jar 文件复制到 Web 应用程序的/WEB-INF/lib 文件夹中。

(2) 加载驱动程序

```
Class.forName("com.microsoft.sqlserver.jdbc.SQLServerDriver");
```

(3) 建立连接

```
Connection con=
    DriverManager.getConnection("jdbc:sqlserver://localhost:1433; DatabaseName=
```

mydatabase ",
    "用户名","密码");
```

### 2. 连接 MySQL 5.5

（1）获取纯 Java 数据库驱动程序

可以登录 MySQL 的官方网站 http://www.mysql.com 下载驱动程序。我们下载的是 mysql-connector-java-5.0.6-bin.jar。然后，把 mysql-connector-java-5.0.6-bin.jar 文件复制到 Web 应用程序的/WEB-INF/lib 文件夹中。

（2）加载驱动程序

```
Class.forName("com.mysql.jdbc.Driver");
```

（3）建立连接

```
Connection con =
    DriverManager.getConnection("jdbc:mysql://localhost:3306/mydatabase","用户名","密码");
```

5.7.2 能力目标

理解使用纯 Java 数据库驱动程序连接不同类型数据库的原理。

5.7.3 任务驱动

1. 任务的主要内容

- 安装 MySQL 5.5；
- 创建数据库和数据表；
- 连接数据库并操作数据表。

2. 任务模板

请按下列步骤操作。

（1）安装 MySQL 5.5

登录官网下载 MySQL 5.5 安装程序 mysql-5.5.19-win32.msi，按照默认安装即可。用户名和密码都设置为 root。

（2）创建数据库和数据表

MySQL 5.5 安装后，选择"开始"→"程序"→MySQL→MySQL Server 5.5→MySQL 5.5 Command Line Client 命令。输入默认密码 root，成功启动 MySQL 监视器后，在 MS-DOS 窗口中出现 mysql>字样，如图 5.24 所示。

成功启动 MySQL 监视器后，就可以使用如下命令。

```
create database mydatabase;
```

创建数据库 mydatabase，为了在 mydatabase 数据库中创建表，必须先进入该数据库，命令如下：

```
use mydatabase
```
```

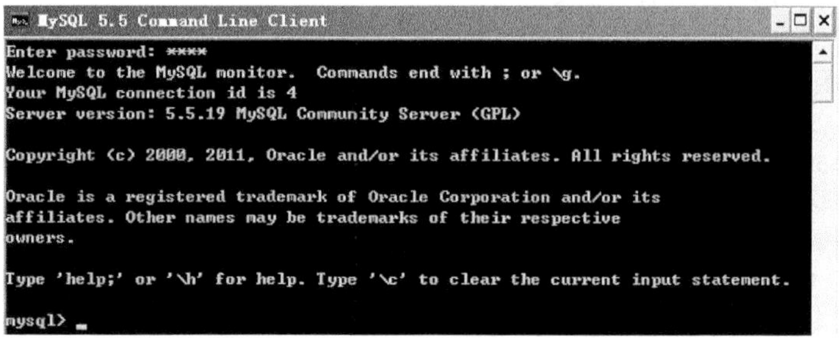

第 5 章 JSP 访问数据库

图 5.24　启动 MySQL 监视器

进入数据库 mydatabase 的操作如图 5.25 所示。

进入数据库后，就可以创建表 employee。

创建表的命令如下：

```
create table employee (
    empNo varchar(10) not null,
    name varchar(30) not null,
    salary float not null,
    primary key (empNo)
);
insert into employee values('001','zhou',1200);
insert into employee values('002','wu',2500);
insert into employee values('003','zheng',2800 );
insert into employee values('004','wang',4800);
insert into employee values('005','zhao',1800);
insert into employee values('006','qian',7800);
commit;
```

（3）连接数据库并操作数据表

编写一个 JSP 页面 useMySQL.jsp，在该页面的 Java 程序片中连接 MySQL 数据库，并从 employee 表中查询出工资大于 2000 的雇员。页面运行效果如图 5.26 所示。

图 5.25　进入数据库

图 5.26　工资大于 2000 的雇员

（4）任务的代码模板

请将 useMySQL.jsp 中的【代码】替换为真正的 JSP 的代码。

useMySQL.jsp

```
<%@page language="java" contentType="text/html; charset=GBK"pageEncoding="GBK"%>
<%@page import="java.sql.*"%>
<html>
<head>
```

```
<title>useMySQL.jsp</title>
</head>
<body bgcolor="lightblue">
    <%
        Connection con=null;
        Statement st=null;
        ResultSet rs=null;
        try {
            【代码 1】                          //加载 MySQL 的 Java 驱动
        } catch (ClassNotFoundException e) {
            e.printStackTrace();
        }
        try {
            【代码 2】    //与 MySQL 建立连接,数据库名为 mydatabase,用户名和密码都为 root
            st=con.createStatement();
            rs=st.executeQuery("select * from employee where salary>2000");
            out.print("<table border=1>");
            out.print("<tr>");
                out.print("<th>雇员编号</th>");
                out.print("<th>雇员姓名</th>");
                out.print("<th>工资</th>");
            out.print("</tr>");
            while(rs.next()){
                out.print("<tr>");
                    out.print("<td>"+rs.getString(1)+"</td>");
                    out.print("<td>"+rs.getString(2)+"</td>");
                    out.print("<td>"+rs.getString(3)+"</td>");
                out.print("</tr>");
            }
            out.print("</table>");
        } catch (SQLException e) {
            e.printStackTrace();
        }finally{
            try{
                if(rs!=null){
                    rs.close();
                }
                if(st!=null){
                    st.close();
                }
                if(con!=null){
                    con.close();
                }
            }catch (SQLException e) {
                e.printStackTrace();
            }
        }
    %>
</body>
</html>
```

3. 任务小结或知识扩展

使用纯 Java 数据库驱动程序连接数据库使得 JSP 程序不依赖具体的操作系统,有利于代码的移植。

4. 代码模板的参考答案

【代码 1】: `Class.forName("com.mysql.jdbc.Driver");`
【代码 2】: `con= DriverManager. getConnection ( " jdbc: mysql://localhost: 3306/mydatabase", "root", "root");`

5.7.4 实践环节

参考本节任务中的主要内容,使用 MySQL 创建一个数据库 yourdatabase,在该数据库中创建一张表 student,并编写程序操作该表。

5.8 PreparedStatement 的使用

与 Statement 语句一样,PreparedStatement 同样可以完成向数据库发送 SQL 语句,获取数据库操作结果的功能。但是 Statement 对象在每次执行 SQL 语句时都将该语句传送给数据库,然后数据库解释器负责将 SQL 语句转换成内部命令,并执行该命令,完成相应的数据库操作。这种机制每次向数据库发送一条 SQL 语句时,都要先转化成内部命令,如果不断地执行程序,就会加重解释器的负担,影响执行的速度。

而 PreparedStatement 对象将 SQL 语句传送给数据库进行预编译,以后需要执行同一条语句时就不需要再重新编译,直接执行就可以了,这样就大大提高了数据库的执行速度。

5.8.1 核心知识

可以使用 Connection 的对象 con 调用 prepareStatement(String sql)方法对参数 sql 指定的 SQL 语句进行预先编译,生成数据库的底层命令,并将该命令封装在 PreparedStatement 对象中。对于 SQL 语句中会变动的部分,可以使用通配符"?"代替。例如:

```
PreparedStatement ps = con.prepareStatement ( "insert into goodsInfo values
(?,?,?,?)";
```

然后使用对应的 setXXX(int parameterIndex,XXX value)方法指定"?"代表的值,其中参数 parameterIndex 用来表示 SQL 语句中从左到右的第 parameterIndex 个通配符号,value 代表该通配符所代表的具体值。例如:

```
ps.setInt(1,9);
ps.setString(2,"手机");
ps.setDouble(3,1900.8);
ps.setString(4,"通信");
```

若要让 SQL 语句执行生效,需使用 PreparedStatement 的对象 ps 调用 executeUpdate()方法。如果是查询,ps 就调用 executeQuery()方法,并返回 ResultSet 对象。

5.8.2 能力目标

能够灵活使用预处理语句对象操作数据库中的表。

5.8.3 任务驱动

1. 任务的主要内容

编写两个 JSP 页面：inputPrepareGoods.jsp 和 showPrepareGoods.jsp。用户可以在 inputPrepareGoods.jsp 页面中输入信息后，单击"添加"按钮把信息添加到 goodsInfo 表中。然后，在 showPrepareGoods.jsp 页面中显示所有商品信息。在该任务中需要编写一个 bean(UsePrepare.java)，bean 中使用预处理语句向 goodsInfo 表中添加记录。页面运行效果如图 5.27(a)、图 5.27(b)所示。

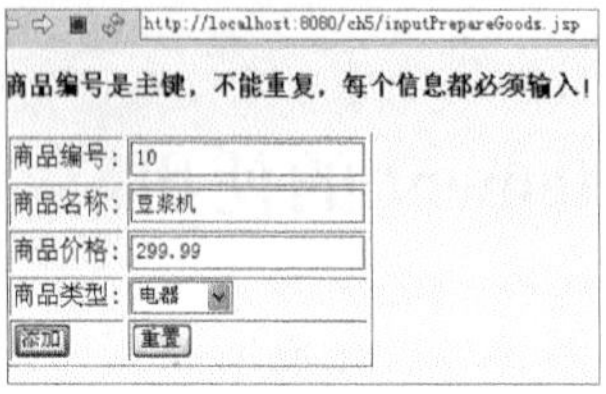

(a) 使用预处理添加商品 　　　　(b) 使用预处理查询商品

图 5.27　5.8 节任务的页面运行效果

2. 任务的代码模板

将下列 UsePrepare.java 中的【代码】替换为真正的 Java 的代码。

inputPrepareGoods.jsp

```
<%@page language="java" contentType="text/html; charset=GBK" pageEncoding="GBK"%>
<html>
<head>
<title>使用预处理语句</title>
</head>
<body bgcolor="LightYellow">
    <h4>商品编号是主键,不能重复,每个信息都必须输入!</h4>
    <form action="showPrepareGoods.jsp" method="post">
    <table border="1">
        <tr>
            <td>商品编号:</td>
            <td><input type="text" name="goodsId"/></td>
        </tr>

        <tr>
            <td>商品名称:</td>
            <td><input type="text" name="goodsName"/></td>
        </tr>

        <tr>
```

```html
        <td>商品价格:</td>
            <td><input type="text" name="goodsPrice"/></td>
        </tr>

        <tr>
            <td>商品类型:</td>
            <td>
                <select name="goodsType">
                    <option value="日用品">日用品
                    <option value="电器">电器
                    <option value="食品">食品
                    <option value="水果">水果
                    <option value="服装">服装
                    <option value="文具">文具
                    <option value="其他">其他
                </select>
            </td>
        </tr>

        <tr>
            <td><input type="submit" value="添加"></td>
            <td><input type="reset" value="重置"></td>
        </tr>
    </table>
    </form>
</body>
</html>
```

showPrepareGoods.jsp

```jsp
<%@page language="java" contentType="text/html; charset=GBK" pageEncoding="GBK"%>
<%@page import="bean.UsePrepare" %>
<html>
<head>
<title>使用预处理语句</title>
</head>
<body>
    <%
        request.setCharacterEncoding("GBK");
    %>
    <jsp:useBean id="prepareGoods" class="bean.UsePrepare" scope="page"></jsp:useBean>
    <jsp:setProperty property="*" name="prepareGoods"/>
    <%
        prepareGoods.addGoods();                        //添加商品
    %>
    <jsp:getProperty property="queryResult" name="prepareGoods"/><!--获得查询结果 -->
</body>
</html>
```

UsePrepare.java

```java
package bean;
import java.sql.*;
public class UsePrepare {
    int goodsId;
    String goodsName;
    double goodsPrice;
    String goodsType;
    StringBuffer queryResult;                          //查询所有商品结果
    public UsePrepare(){

    }
    public int getGoodsId() {
        return goodsId;
    }
    public void setGoodsId(int goodsId) {
        this.goodsId=goodsId;
    }
    public String getGoodsName() {
        return goodsName;
    }
    public void setGoodsName(String goodsName) {
        this.goodsName=goodsName;
    }
    public double getGoodsPrice() {
        return goodsPrice;
    }
    public void setGoodsPrice(double goodsPrice) {
        this.goodsPrice=goodsPrice;
    }
    public String getGoodsType() {
        return goodsType;
    }
    public void setGoodsType(String goodsType) {
        this.goodsType=goodsType;
    }
    //添加商品
    public void addGoods(){
    Connection con=null;
        PreparedStatement ps=null;
        try {
            Class.forName("oracle.jdbc.driver.OracleDriver");
        } catch (ClassNotFoundException e) {
            e.printStackTrace();
        }
        try {
            con= DriverManager.getConnection("jdbc:oracle:thin:@localhost:1521:orcl",
                    "system","system");
            ps=con.prepareStatement("insert into goodsInfo values(?,?,?,?)");
```

```
            【代码 1】            //ps 调用 set 方法指定第 1 个通配符的值
            【代码 2】            //ps 调用 set 方法指定第 2 个通配符的值
            【代码 3】            //ps 调用 set 方法指定第 3 个通配符的值
            【代码 4】            //ps 调用 set 方法指定第 4 个通配符的值
            ps.executeUpdate();
        }catch (SQLException e) {
            e.printStackTrace();
        }finally{
            try{
                if(ps!=null){
                    ps.close();
                }
                if(con!=null){
                    con.close();
                }
            }catch (SQLException e) {
                e.printStackTrace();
            }
        }
    }
    //获得所有商品信息
    public StringBuffer getQueryResult(){
        queryResult=new StringBuffer();
        Connection con=null;
        PreparedStatement ps=null;
        ResultSet rs=null;
        try {
            Class.forName("oracle.jdbc.driver.OracleDriver");
        } catch (ClassNotFoundException e) {
            e.printStackTrace();
        }
        try {
             con= DriverManager.getConnection ("jdbc:oracle:thin:@ localhost:
             1521:orcl",
                    "system","system");
            ps=con.prepareStatement("select * from goodsInfo");
            rs=ps.executeQuery();
            queryResult.append("<table border=1>");
                queryResult.append("<tr>");
                    queryResult.append("<th>goodsId</th>");
                    queryResult.append("<th>goodsName</th>");
                    queryResult.append("<th>goodsPrice</th>");
                    queryResult.append("<th>goodsType</th>");
                queryResult.append("</tr>");
            while(rs.next()){
                queryResult.append("<tr>");
                    queryResult.append("<td>"+rs.getString(1)+"</td>");
                    queryResult.append("<td>"+rs.getString(2)+"</td>");
                    queryResult.append("<td>"+rs.getString(3)+"</td>");
                    queryResult.append("<td>"+rs.getString(4)+"</td>");
                queryResult.append("</tr>");
```

```
            }
            queryResult.append("</table>");
        }catch (SQLException e) {
            e.printStackTrace();
        }finally{
            try{
                if(rs!=null){
                    rs.close();
                }
                if(ps!=null){
                    ps.close();
                }
                if(con!=null){
                    con.close();
                }
            }catch (SQLException e) {
                e.printStackTrace();
            }
        }
        return queryResult;
    }
}
```

3. 任务小结或知识扩展

Statement 在执行 executeQuery(String sql)、executeUpdate(String sql) 等方法时，如果 SQL 语句有些部分是动态的数据，必须使用 "+" 连字符组成完整的 SQL 语句，十分不便。例如，5.3 节中的任务在添加商品时，必须按如下方式组成 SQL 语句。

```
String addSql="insert into goodsInfo values("+goodsId+",'"+goodsName+"','"+goodsPrice+",'"+goodsType+"')";
st.executeUpdate(addSql);
```

PreparedStatement 对象被称为预处理语句对象，现在使用预处理语句不仅提高了数据库的访问效率，而且方便了程序的编写。预处理语句对象调用 executeUpdate() 和 executeQuery() 方法时不需要传递参数。例如：

```
int i=ps.executeUpdate();
```

或

```
ResultSet rs=ps.executeQuery();
```

4. 代码模板的参考答案

【代码 1】：`ps.setInt(1, goodsId);`
【代码 2】：`ps.setString(2, goodsName);`
【代码 3】：`ps.setDouble(3, goodsPrice);`
【代码 4】：`ps.setString(4, goodsType);`

5.8.4 实践环节

编写两个 JSP 页面：inputPrepareQuery.jsp 和 showPrepareBy.jsp。用户先可以在页面 inputPrepareQuery.jsp 中输入查询条件，再单击"查询"按钮。然后，在 showPrepareBy.jsp 页面中显示符合查询条件的商品信息。在本节任务的 bean（UsePrepare.java）中添加一个方法 getQueryPrepareResultBy() 实现该题的条件查询功能（使用预处理语句实现查询）。页面运行效果如图 5.28(a)、图 5.28(b) 所示。

(a) 使用预处理条件查询　　　　(b) 符合查询条件的记录

图 5.28　5.8 节实践环节页面运行效果图

5.9 小　　结

- JDBC 连接数据库有两种常用方式：建立 JDBC-ODBC 桥接器和加载纯 Java 驱动程序。
- 目前有多种类型的数据库。JDBC 不管连接哪种类型的数据库，连接方式基本上是类似的。需要重点关注的是，在连接各数据库时驱动程序加载部分的代码和连接部分的代码。
- 数据库连接池的基本思想是：为数据库连接建立一个"缓冲池"。预先在"缓冲池"中放入一定数量的连接，当需要建立数据库连接时，只需从"缓冲池"中取出一个，使用完毕之后再放回去。可以通过设定连接池最大连接数来防止系统无限度地与数据库连接。更为重要的是，通过连接池的管理机制监视数据库连接的数量及使用情况，为系统开发、测试和性能调整提供依据。
- 预处理语句对象不仅大大提高了数据库的执行速度，而且方便了程序的编写。

习　题　5

1. 当在 JSP 文件中要编写代码连接数据库时，应在 JSP 文件中加入以下哪个语句？（　　）

    A. <jsp:include file="java.util.*"/>

    B. <%@page import="java.sql.*"%>

    C. <jsp:include page="java.lang.*"/>

    D. <%@page import="java.util.*"%>

2. Java 程序连接数据库常用的两种方式是：建立 JDBC-ODBC 桥接器和加载纯（　　）

驱动程序。

    A. Oracle   B. Java   C. Java 数据库   D. 以上都不对

3. 从"员工"表的"姓名"字段找出名字包含"玛丽"的人，下面哪条 SELECT 语句正确？（　　）

    A. select * from 员工 where 姓名 like'％玛丽％'

    B. select * from 员工 where 姓名＝'％玛丽_'

    C. select * from 员工 where 姓名 like'_玛丽％'

    D. select * from 员工 where 姓名＝'_玛丽_'

4. 下述选项中不属于 JDBC 基本功能的是（　　）。

    A. 与数据库建立连接   B. 提交 SQL 语句

    C. 处理查询结果   D. 数据库维护管理

5. 请选出微软公司提供的连接 SQL Server 2005 的 JDBC 驱动程序。（　　）

    A. com.mysql.jdbc.Driver

    B. sun.jdbc.odbc.JdbcOdbcDriver

    C. oracle.jdbc.driver.OracleDriver

    D. com.microsoft.sqlserver.jdbc.SQLServerDriver

6. 下面（　　）不是 ResultSet 接口的方法。

    A. next()   B. getString()   C. back()   D. getInt()

7. JDBC 能完成哪些工作？

8. 使用纯 Java 数据库驱动程序访问数据库时，有哪些步骤？

9. JDBC 连接数据库常用的方式有哪些？

Java Servlet 基础

本章主要内容

- Servlet 类与 servlet 对象
- servlet 对象的创建与运行
- 通过 JSP 页面访问 servlet
- doGet 和 doPost 方法
- 重定向与转发
- 会话管理

Java Servlet 的核心思想就是在 Web 服务器端创建用来响应用户请求的对象,该对象被称为一个 servlet 对象。JSP 技术以 Java Servlet 为基础,当客户请求一个 JSP 页面时,Web 服务器(如 Tomcat 服务器)会自动生成一个对应的 Java 文件,编译该 Java 文件,并用编译得到的字节码文件在服务器端创建一个 servlet 对象。但大多数 Web 应用需要 servlet 对象具有特定的功能,这时就需要 Web 开发人员自己编写创建 servlet 对象的类。如何编写 Servlet 类,又如何使用 Servlet 类,关于这些将在本章重点介绍。

6.1 Servlet 类与 servlet 对象

6.1.1 核心知识

编写一个 Servlet 类很简单,只要继承 javax.servlet.http 包中的 HttpServlet 类,并重写响应 HTTP 请求的方法即可。HttpServlet 类实现了 Servlet 接口,实现了响应用户的接口方法。HttpServlet 类的一个子类习惯上称为一个 Servlet 类,这样的子类创建的对象又习惯地称为 servlet 对象。

6.1.2 能力目标

理解 Servlet 类与 servlet 对象的概念。

6.1.3 任务驱动

1. 任务的主要内容

编写一个简单的 Servlet 类 MyFirstServlet,用户请求这个 servlet 对象时,就会在浏览器中看到"人生第一个 Servlet 类"这样的响应信息。

2. 任务的代码模板

请将下列 MyFirstServlet.java 中的【代码】替换为 Java 代码。

MyFirstServlet.java

```java
package servlet;
import java.io.*;
import javax.servlet.*;
import javax.servlet.http.*;
public class MyFirstServlet extends 【代码】{
    public void init(ServletConfig config) throws ServletException{
        super.init(config);
    }
    public void service(HttpServletRequest request,HttpServletResponse response)
            throws IOException{
        //设置响应的内容类型
        response.setContentType("text/html;charset=utf-8");
        //取得输出对象
        PrintWriter out=response.getWriter();
        out.println("<html><body>");
        //在浏览器中显示:人生第一个 Servlet 类
        out.println("人生第一个 Servlet 类");
        out.println("</body></html>");
    }
}
```

3. 任务小结或知识扩展

编写 Servlet 类时,必须有包名。也就是说,必须在包中编写 Servlet 类。在本章中我们新建一个 Web 工程 ch6,所有的 Servlet 类放在 src 下 servlet 目录中。

任务中 Servlet 类的源文件 MyFirstServlet.java 保存在 Eclipse 的 Web 工程 ch6 的 src 下 servlet 目录中。MyFirstServlet.java 源文件由 Eclipse 自动编译生成字节码文件 MyFirstServlet.class,保存在 build\classes\servlet 里面。Servlet 类的源文件与字节码文件保存目录如图 6.1 所示。

图 6.1 Servlet 类的源文件与字节码文件保存目录

编写完 Servlet 类的源文件,并编译了源文件,这时是不是就可以运行 servlet 对象呢?不可以,需要部署 servlet 之后,才可以运行 servlet 对象(见后面的 6.2 节)。

4. 代码模板的参考答案

【代码】:HttpServlet

6.1.4　实践环节

编写一个简单的 Servlet 类 YourFirstServlet。用户请求这个 servlet 对象时,就会在浏览器中看到"您人生中第一个 Servlet 类"这样的响应信息,并在您的 Web 工程中找到该 Servlet 类对应的字节码文件。

6.2　servlet 对象的创建与运行

6.2.1　核心知识

要想让 Web 服务器用 Servlet 类编译后的字节码文件创建 servlet 对象的话,必须为 Web 服务器编写一个部署文件。这个部署文件是一个 XML 文件,名字是 web.xml,该文件由 Web 服务器负责管理。在 ch6\WebContent\WEB-INF 目录里找到 web.xml 文件,可以在该文件里部署自己的 servlet。

6.2.2　能力目标

掌握部署与运行 servlet 的方法。

6.2.3　任务驱动

1. 任务的主要内容
- 部署 servlet;
- 运行 servlet。

2. 任务模板

按下列步骤操作。

(1) 部署 servlet

为了在 web.xml 文件里部署 6.1 节中的 MyFirstServlet,需要在 web.xml 文件里找到＜web-app＞＜/web-app＞标记,然后在＜web-app＞＜/web-app＞标记中添加如下内容。

```
<servlet>
    <servlet-name>myServlet</servlet-name>
    <servlet-class>servlet.MyFirstServlet</servlet-class>
</servlet>
<servlet-mapping>
    <servlet-name>myServlet</servlet-name>
    <url-pattern>/firstServlet</url-pattern>
</servlet-mapping>
```

(2) 运行 servlet

在 web.xml 文件中部署完 servlet 之后,就可以运行 servlet 了。servlet 第一次被直接访问(可以通过 JSP 页面间接访问)时,需要把它发布到 Web 服务器上(选中 Servlet 类的源文件后右击,选择 Run As→Run On Server 命令)。这时,在 Eclipse 内嵌的浏览器中看到如图 6.2 所示的画面。

把 servlet 发布到 Web 服务器上之后,也可以在 IE 浏览器的地址栏中输入:

http://localhost:8080/ch6/firstServlet

图 6.2 servlet 运行效果

来请求运行 servlet。

3. 任务小结或知识扩展

1) web.xml 文件中的具体内容及其作用

(1) 根标记<web-app>

XML 文件中必须有一个根标记,web.xml 的根标记是<web-app>。

(2) <servlet>标记及其子标记

web.xml 文件中可以有若干个<servlet>标记,该标记的内容由 Web 服务器负责处理。<servlet>标记中有<servlet-name>和<servlet-class>两个子标记,其中<servlet-name>子标记的内容是 Web 服务器创建的 servlet 对象的名字。web.xml 文件中可以有若干个<servlet>标记,但要求它们的<servlet-name>子标记的内容互不相同。<servlet-class>子标记的内容指定 Web 服务器用哪个类来创建 servlet 对象,如果 servlet 对象已经创建,那么 Web 服务器就不再使用指定的类创建。

(3) <servlet-mapping>标记及其子标记

web.xml 文件中出现一个<servlet>标记就会对应地出现一个<servlet-mapping>标记。<servlet-mapping>标记中也有两个子标记:<servlet-name>和<url-pattern>。其中<servlet-name>子标记的内容是 Web 服务器创建的 servlet 对象的名字(该名字必须和<servlet>标记的子标记<servlet-name>的内容相同);<url-pattern>子标记用来指定用户用怎样的模式请求 servlet 对象,比如,<url-pattern>子标记的内容是/firstServlet,用户要请求服务器运行 servlet 对象 myServlet 为其服务,那么在 IE 浏览器的地址栏中输入:

http://localhost:8080/ch8/firstServlet

一个 Web 服务器的 web.xml 文件负责管理该 Web 服务的 servlet 对象,当 Web 服务需要提供更多的 servlet 对象时,只要在 web.xml 文件中添加<servlet>和<servlet-mapping>标记即可。

2) 关于 servlet 的生命周期

一个 servlet 对象的生命周期主要由下列三个过程组成。

(1) 初始化 servlet 对象

当 servlet 对象第一次被请求加载时,服务器会创建一个 servlet 对象,该 servlet 对象调用 init 方法完成必要的初始化工作。

(2) service 方法响应请求

创建的 servlet 对象再调用 service 方法响应客户的请求。

(3) servlet 对象死亡

当服务器关闭时,servlet 对象调用 destroy 方法使自己消亡。

从上面三个过程来看，init 方法只能被调用一次，即在 servlet 第一次被请求加载时调用该方法。当客户请求 servlet 服务时，服务器将启动一个新的线程，在该线程中，servlet 对象调用 service 方法响应客户的请求。那么多个客户请求 servlet 服务时，服务器会怎么办呢？服务器会为每个客户启动一个新的线程，在每个线程中，servlet 对象调用 service 方法响应客户的请求。也就是说，每个客户请求都会导致 service 方法被调用执行，分别运行在不同的线程中。

3）Servlet 类中的方法

(1) init 方法

init 方法是 HttpServlet 类中的方法，可以在 Servlet 类中被重写。init 方法的声明格式如下：

```
public void init(ServletConfig config) throws ServletException
```

servlet 对象第一次被请求加载时，服务器创建一个 servlet 对象，该对象调用 init 方法完成必要的初始化工作。

(2) service 方法

service 方法是 HttpServlet 类中的方法，可以在 Servlet 类中被重写。service 方法的声明格式如下：

```
public void service(HttpServletRequest request,HttpServletResponse response)
throws IOException
```

当 servlet 对象成功创建后，servlet 对象就调用 service 方法来处理用户的请求并返回响应。与 init 方法不同的是，init 方法只能被调用一次，而 service 方法可能被多次调用。我们知道，当客户请求 servlet 对象时，服务器将启动一个新的线程，在该线程中，servlet 对象调用 service 方法响应客户的请求。换句话说，每个客户的每次请求都导致 service 方法被调用执行，调用过程运行在不同的线程中，互不干扰。

(3) destroy 方法

destroy 方法是 HttpServlet 类中的方法。一个 Servlet 类可直接继承该方法，一般不需要重写。destroy 方法的声明格式如下：

```
public void destroy()
```

当服务器终止服务时，destroy 方法会被执行，使 servlet 对象消亡。

4. 代码模板的参考答案

无参考答案

6.2.4 实践环节

编写一个简单的 Servlet 类 Pratice2Servlet，在 web.xml 中部署该 servlet 并运行它。用户通过在 IE 浏览器地址栏中输入"http://localhost:8080/ch8/pratice2"请求这个 servlet 对象时，就会在浏览器中看到"部署与运行 servlet"这样的响应信息。

6.3 通过 JSP 页面访问 servlet

可以通过 JSP 页面的表单或超链接请求某个 servlet。通过 JSP 页面访问 servlet 的好处是：JSP 页面负责页面的静态信息处理，动态信息处理由 servlet 完成。本章所涉及的 JSP 页面均保存在 ch6\WebContent 目录里。

6.3.1 核心知识

1. 通过表单访问 servlet

假设在 JSP 页面中，有如下表单。

```
<form action="isLogin" method="post">
    ...
</form>
```

那么该表单的处理程序（action）就是一个 servlet，为该 servlet 部署时，web.xml 文件中的标记<servlet-mapping>的子标记<url-pattern>的内容是"/isLogin"。

2. 通过超链接访问 servlet

在 JSP 页面中，可以单击超链接，访问 servlet 对象，同时也可以通过超链接向 servlet 提交信息，例如，<a href="isLogin? user＝kazhafei＆＆pwd＝aobama">查看用户名和密码</a>，"查看用户名和密码"这个超链接就把 user＝kazhafei 和 pwd＝aobama 两个信息提交给了 servlet 处理。

6.3.2 能力目标

能够灵活使用 JSP 页面访问 servlet 对象。

6.3.3 任务驱动

1. 任务的主要内容

编写一个 JSP 页面 login.jsp，在该页面中通过表单向名字为 login 的 servlet 对象（由 LoginServlet 类负责创建）提交用户名和密码，servlet 负责判断输入的用户名和密码是否正确，并把判断结果返回给客户。需要为 web.xml 文件添加如下的子标记。

```
<servlet>
    <servlet-name>login</servlet-name>
    <servlet-class>servlet.LoginServlet</servlet-class>
</servlet>
<servlet-mapping>
    <servlet-name>login</servlet-name>
    <url-pattern>/isLogin</url-pattern>
</servlet-mapping>
```

页面运行效果如图 6.3(a)～图 6.3(c)所示。

2. 任务的代码模板

将下列 login.jsp 页面中的【代码】替换为 JSP 的代码。

第 6 章 Java Servlet 基础

(a) 信息输入页面

(b) 错误信息   (c) 正确信息

图 6.3   6.3 节任务的页面运行效果图

login.jsp

```
<%@page language="java" contentType="text/html; charset=GBK" pageEncoding="GBK"%>
<html>
  <head>
    <title>login.jsp</title>
  </head>
  <body>
    <form【代码】 method="post">
      <table>
        <tr>
          <td>用户名:</td>
          <td><input type="text" name="user"/></td>
        </tr>
        <tr>
          <td>密  码:</td>
          <td><input type="password" name="pwd"/></td>
        </tr>
        <tr>
          <td><input type="submit" value="提交"/></td>
          <td><input type="reset" value="重置"/></td>
        </tr>
      </table>
    </form>
  </body>
</html>
```

LoginServlet.java

```
package servlet;
import java.io.*;
import javax.servlet.*;
import javax.servlet.http.*;
public class LoginServlet extends HttpServlet {
    public void init(ServletConfig config) throws ServletException{
        super.init(config);
    }
```

```java
public void service(HttpServletRequest request,HttpServletResponse response)
    throws IOException{
    response.setContentType("text/html;charset=GBK");
    PrintWriter out=response.getWriter();
    request.setCharacterEncoding("GBK");            //设置编码,防止中文乱码
    String name=request.getParameter("user");       //获取客户提交的信息
    String password=request.getParameter("pwd");    //获取客户提交的信息
    out.println("<html><body>");
    if(name==null||name.length()==0){
        out.println("请输入用户名");
    }
    else if(password==null||password.length()==0){
        out.println("请输入密码");
    }
    else if(name.length()>0&&password.length()>0){
        if(name.equals("kazhafei")&&password.equals("aobama")){
            out.println("信息输入正确");
        }else{
            out.println("信息输入错误");
        }
    }
    out.println("</body></html>");
    }
}
```

3. 任务小结或知识扩展

需要特别注意的是,如果 servlet 的请求格式是"/XXX"(请求格式就是 web.xml 文件中的标记＜servlet-mapping＞的子标记＜url-pattern＞的内容),那么 JSP 页面请求 servlet 时,必须写成"XXX",不可以写成"/XXX",否则将变成请求服务器(Tomcat)root 目录下的某个 servlet。

4. 代码模板的参考答案

【代码】: action="isLogin"

6.3.4 实践环节

把例任务中 login.jsp 的信息提交方式(表单提交)改成超链接提交方式来访问 servlet。提示:超链接访问 servlet 格式如下:

```
<a href="isLogin? user=kazhafei&&pwd=aobama">查看用户名和密码</a>
```

6.4 doGet 和 doPost 方法

我们编写 Servlet 类时,经常重写 HttpServlet 类中的 doGet 和 doPost 方法,用来处理客户的请求并作出响应。

6.4.1 核心知识

当服务器接收到一个 servlet 请求时,就会产生一个新线程,在这个线程中让 servlet 对象调用 service 方法为请求作出响应。service 方法首先检查 HTTP 请求类型(get 或 post),并在 service 方法中根据用户的请求类型,对应地再调用 doGet 或 doPost 方法。

HTTP 请求类型为 get 方式时,service 方法调用 doGet 方法响应用户请求;HTTP 请求类型为 post 方式时,service 方法调用 doPost 方法响应用户请求。因此,在 Servlet 类中,没有必要重写 service 方法,直接继承即可。

在 Servlet 类中重写 doGet 或 doPost 方法来响应用户的请求,这样可以增加响应的灵活性,减轻服务器的负担。

6.4.2 能力目标

理解 doGet 和 doPost 方法的调用原理。

6.4.3 任务驱动

1. 任务的主要内容

编写一个 JSP 页面 input.jsp,在该页面中使用表单向 servlet 对象 computer 提交矩形的长与宽的值。computer(由 GetLengthOrAreaServlet 负责创建)处理手段依赖表单提交数据的方式,当提交方式为 get 时,computer 对象计算矩形的周长;当提交方式为 post 时,computer 对象计算矩形的面积。需要为 web.xml 文件添加如下的子标记。

```xml
<servlet>
    <servlet-name>computer</servlet-name>
    <servlet-class>servlet.GetLengthOrAreaServlet</servlet-class>
</servlet>
<servlet-mapping>
    <servlet-name>computer</servlet-name>
    <url-pattern>/showLengthOrArea</url-pattern>
</servlet-mapping>
```

页面运行效果如图 6.4(a)~图 6.4(c)所示。

2. 任务的代码模板

将下列 input.jsp 页面中的【代码】替换为 JSP 的代码。

input.jsp

```jsp
<%@page language="java" contentType="text/html; charset=GBK" pageEncoding="GBK"%>
<html>
    <head>
        <title>input.jsp</title>
    </head>
    <body>
        <h2>输入矩形的长和宽,提交给 servlet(post 方式)求面积:</h2>
```

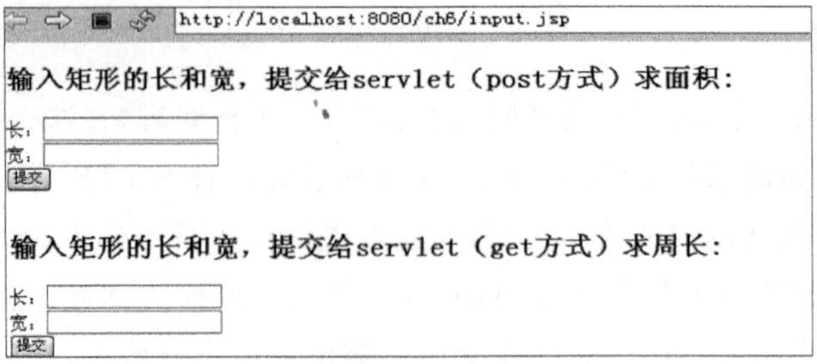

(a) 信息输入页面

(b) post方式提交获得矩形面积

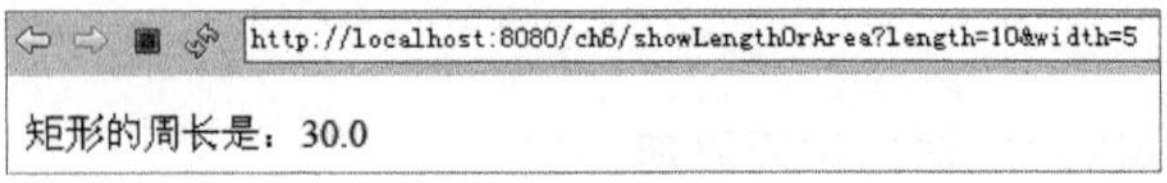

(c) get方式提交获得矩形周长

图 6.4　6.4节任务的页面运行效果图

```
    <form【代码 1】>
    长:<input type="text" name="length"/><br/>
    宽:<input type="text" name="width"/><br/>
    <input type="submit" value="提交"/>
    </form>
    <br/>
    <h2>输入矩形的长和宽,提交给 servlet(get方式)求周长:</h2>
    <form【代码 2】>
    长:<input type="text" name="length"/><br/>
    宽:<input type="text" name="width"/><br/>
    <input type="submit" value="提交"/>
    </form>
  </body>
</html>
```

GetLengthOrAreaServlet.java

```
package servlet;
import java.io.*;
import javax.servlet.*;
import javax.servlet.http.*;
public class GetLengthOrAreaServlet  extends HttpServlet {
    public void init(ServletConfig config) throws ServletException{
        super.init(config);
    }
    public void doPost(HttpServletRequest request,HttpServletResponse
            response) throws ServletException,IOException{
        response.setContentType("text/html;charset=GBK");
        PrintWriter out=response.getWriter();
        String l=request.getParameter("length");
        String w=request.getParameter("width");
```

```
            out.println("<html><body>");
            double m=0,n=0;
            try{
                m=Double.parseDouble(l);
                n=Double.parseDouble(w);
                【代码3】                    //计算矩形的面积,并在浏览器上输出
            }catch(NumberFormatException e){
                out.println("请输入数字字符!");
            }
            out.println("</body></html>");
        }
        public void doGet(HttpServletRequest request,HttpServletResponse
                response) throws ServletException,IOException{
            response.setContentType("text/html;charset=utf-8");
            PrintWriter out=response.getWriter();
            String l=request.getParameter("length");
            String w=request.getParameter("width");
            out.println("<html><body>");
            double m=0,n=0;
            try{
                m=Double.parseDouble(l);
                n=Double.parseDouble(w);
                【代码4】                    //计算矩形的周长,并在浏览器上输出
            }catch(NumberFormatException e){
                out.println("请输入数字字符!");
            }
            out.println("</body></html>");
        }
    }
```

3. 任务小结或知识扩展

一般情况下,如果不论用户请求类型是 get 还是 post,服务器的处理过程完全相同,那么可以只在 doPost 方法中编写处理过程,而在 doGet 方法中再调用 doPost 方法,或只在 doGet 方法中编写处理过程,而在 doPost 方法中再调用 doGet 方法。

4. 代码模板的参考答案

【代码1】:action="showLengthOrArea" method="post"
【代码2】:action="showLengthOrArea" method="get"
【代码3】:out.println("矩形的面积是:"+m*n);
【代码4】:out.println("矩形的周长是:"+(m+n)*2);

6.4.4 实践环节

编写一个 JSP 页面 pratice6_4.jsp,在该 JSP 页面中用户可以使用表单向 servlet 对象 pratice 提交矩形的长与宽的值。pratice(由 PraticeServlet 类负责创建)处理数据的手段不依赖表单提交数据的方式,即不论 post 还是 get,处理数据的手段相同,都是计算矩形的周长。

6.5 重定向与转发

重定向是将用户从当前 JSP 页面或 servlet 定向到另一个 JSP 页面或 servlet，以前的 request 中存放的信息全部失效，并进入一个新的 request 作用域；转发是将用户对当前 JSP 页面或 servlet 的请求转发给另一个 JSP 页面或 servlet，以前的 request 中存放的信息不会失效。

6.5.1 核心知识

1. 重定向

重定向方法 public void sendRedirect(String location) 是 HttpServletResponse 类中的方法。重定向的目标页面或 servlet(由参数 location 指定)无法从以前的 request 对象获取用户提交的数据。

2. 转发

javax.servlet.RequestDispatcher 对象可以把用户对当前 JSP 页面或 servlet 的请求转发给另一个 JSP 页面或 servlet。实现转发需要如下两个步骤。

(1) 获得 RequestDispatcher 对象

在当前 JSP 页面或 servlet 中，使用 request 对象调用

```
public RequestDispatcher getRequestDispatcher(String url)
```

方法返回一个 RequestDispatcher 对象，其中参数 url 就是要转发的 JSP 页面或 servlet 的地址，例如：

```
RequestDispatcher dis=request.getRequestDispatcher("dologin");
```

(2) RequestDispatcher 对象调用 forward 方法实现转发

获得 RequestDispatcher 对象之后，就可以使用该对象调用

```
public void forward(ServletRequest request,ServletResponse response)
```

方法将用户对当前 JSP 页面或 servlet 的请求转发给 RequestDispatcher 对象所指定的 JSP 页面或 servlet。例如：

```
dis.forward(request,response);
```

将用户对当前 JSP 页面或 servlet 的请求转变成对 dologin(servlet)的请求。

6.5.2 能力目标

理解重定向与转发的区别，掌握重定向与转发的实现方法。

6.5.3 任务驱动

1. 任务的主要内容

编写 JSP 页面 redirectForward.jsp，在该 JSP 页面中通过表单向名字为 rforward 的

servlet 对象(由 RedirectForwardServlet 类负责创建)提交用户名和密码。如果用户输入的数据不完整(没有输入用户名或密码),那么 rforward 就将用户重定向到 redirectForward.jsp 页面;如果用户输入的数据完整,rforward 就将用户对 redirectForward.jsp 页面的请求转发给名字为 show 的 servlet 对象(由 ShowServlet 类负责创建),该 servlet 对象显示用户输入的信息。需要为 web.xml 文件添加如下的子标记。

```xml
<servlet>
    <servlet-name>rforward</servlet-name>
    <servlet-class>servlet.RedirectForwardServlet</servlet-class>
</servlet>
<servlet>
    <servlet-name>show</servlet-name>
    <servlet-class>servlet.ShowServlet</servlet-class>
</servlet>
<servlet-mapping>
    <servlet-name>rforward</servlet-name>
    <url-pattern>/rfLogin</url-pattern>
</servlet-mapping>
<servlet-mapping>
    <servlet-name>show</servlet-name>
    <url-pattern>/yourMessage</url-pattern>
</servlet-mapping>
```

2. 任务的代码模板

将下列 RedirectForwardServlet.java 中的【代码】替换为 Java 代码。

redirectForward.jsp

```jsp
<%@ page language="java" contentType="text/html; charset=GBK" pageEncoding="GBK"%>
<html>
    <head>
        <title>redirectForward.jsp</title>
    </head>
    <body>
        <form action="rfLogin" method="post">
            <table>
                <tr>
                    <td>用户名:</td>
                    <td><input type="text" name="user"/></td>
                </tr>
                <tr>
                    <td>密　码:</td>
                    <td><input type="password" name="pwd"/></td>
                </tr>
                <tr>
                    <td><input type="submit" value="提交"/></td>
                    <td><input type="reset" value="重置"/></td>
                </tr>
            </table>
```

```
        </form>
    </body>
</html>
```

RedirectForwardServlet.java

```java
package servlet;
import java.io.IOException;
import javax.servlet.RequestDispatcher;
import javax.servlet.ServletConfig;
import javax.servlet.ServletException;
import javax.servlet.http.HttpServlet;
import javax.servlet.http.HttpServletRequest;
import javax.servlet.http.HttpServletResponse;
public class RedirectForwardServlet extends HttpServlet {
    public void init(ServletConfig config) throws ServletException{
        super.init(config);
    }
    public void doPost(HttpServletRequest request,HttpServletResponse
            response) throws ServletException,IOException{
        String name=request.getParameter("user");
        String password=request.getParameter("pwd");
        if(name==null||name.length()==0){
            【代码 1】
                //使用 response 调用 sendRedirect 方法重定向到 redirectForward.jsp
        }
        else if(password==null||password.length()==0){
            【代码 2】
                //使用 response 调用 sendRedirect 方法重定向到 redirectForward.jsp
        }
        else if(name.length()>0&&password.length()>0){
            【代码 3】//获得 RequestDispatcher 对象 dis,转发到 servlet 对象 yourMessage
            【代码 4】//dis 对象调用 forward 方法实现转发
        }
    }
    public void doGet(HttpServletRequest request,HttpServletResponse
            response) throws ServletException,IOException{
        doPost(request,response);
    }
}
```

ShowServlet.java

```java
package servlet;
import java.io.*;
import javax.servlet.*;
import javax.servlet.http.*;
public class ShowServlet extends HttpServlet {
    public void init(ServletConfig config) throws ServletException{
        super.init(config);
    }
    public void doPost(HttpServletRequest request,HttpServletResponse
```

```
            response) throws ServletException,IOException{
        response.setContentType("text/html;charset=GBK");
        PrintWriter out=response.getWriter();
        String name=request.getParameter("user");
        String password=request.getParameter("pwd");
        byte b[]=name.getBytes("ISO-8859-1");
        name=new String(b,"UTF-8");
        byte b1[]=password.getBytes("ISO-8859-1");
        password=new String(b1,"UTF-8");
        out.println("您的用户名是:"+name);
        out.println("<br>您的密码是:"+password);
    }
    public void doGet(HttpServletRequest request,HttpServletResponse
            response) throws ServletException,IOException{
        doPost(request,response);
    }
}
```

3. 任务小结或知识扩展

和重定向方法不同的是,使用转发时,用户在浏览器的地址栏中不能看到 forward 方法所转发的页面地址或 servlet 的地址,只能看到它们的运行效果。用户在浏览器的地址栏中所看到的仍然是当前 JSP 页面或 servlet 的地址。

4. 代码模板的参考答案

【代码 1】: response.sendRedirect("redirectForward.jsp");
【代码 2】: response.sendRedirect("redirectForward.jsp");
【代码 3】: RequestDispatcher dis= request.getRequestDispatcher("yourMessage");
【代码 4】: dis.forward(request, response);

6.5.4 实践环节

试着把任务中的转发(【代码 3】与【代码 4】部分)改成重定向,然后运行 redirectForward.JSP 页面,看看运行结果是怎样的?为什么是这样的结果?

6.6 在 servlet 中使用 session

6.6.1 核心知识

在 servlet 中可以使用 request 对象调用 getSession 方法获取用户的会话对象 session,例如:

```
HttpSession session= request.getSession(true);
```

一个用户在同一个 Web 服务的不同 servlet 对象中获取的 session 对象是完全相同的,但是不同用户的 session 对象不同。

6.6.2 能力目标

掌握如何在 servlet 中获取用户的会话对象 session。

6.6.3 任务驱动

1. 任务的主要内容

编写一个 JSP 页面 useSession.jsp，在该页面中通过表单向名字为 useSession 的 servlet 对象（由 UseSessionServlet 类负责创建）提交用户名，useSession 将用户名存入用户的 session 对象中，然后用户请求另一个 servlet 对象 showName（由 ShowNameServlet 类负责创建），showName 从用户的 session 对象中取出存储的用户名，并显示在浏览器中。需要为 web.xml 文件添加如下的子标记。

```xml
<servlet>
    <servlet-name>useSession</servlet-name>
    <servlet-class>servlet.UseSessionServlet</servlet-class>
</servlet>
<servlet>
    <servlet-name>showName</servlet-name>
    <servlet-class>servlet.ShowNameServlet</servlet-class>
</servlet>
<servlet-mapping>
    <servlet-name>useSession</servlet-name>
    <url-pattern>/sendMyName</url-pattern>
</servlet-mapping>
<servlet-mapping>
    <servlet-name>showName</servlet-name>
    <url-pattern>/showMyName</url-pattern>
</servlet-mapping>
```

页面运行效果如图 6.5(a)～图 6.5(c)所示。

(a) 信息输入页面 　　　　(b) 获取会话并存储数据

(c) 获取会话中的数据并显示

图 6.5　6.6 节任务的页面运行效果图

2. 任务的代码模板

将下列 ShowNameServlet.java 和 UseSessionServlet.java 中的【代码】替换为 Java 代码。

useSession.jsp

```jsp
<%@page language="java" contentType="text/html; charset=GBK" pageEncoding="GBK"%>
```

```html
<html>
    <head>
        <title>useSession.jsp</title>
    </head>
    <body>
        <form action="sendMyName" method="post">
            <table>
                <tr>
                    <td>用户名:</td>
                    <td><input type="text" name="user"/></td>
                </tr>
                <tr>
                    <td><input type="submit" value="提交"/></td>
                </tr>
            </table>
        </form>
    </body>
</html>
```

UseSessionServlet. java

```java
package servlet;
import java.io.*;
import javax.servlet.*;
import javax.servlet.http.*;
public class UseSessionServlet extends HttpServlet {
    public void init(ServletConfig config) throws ServletException{
        super.init(config);
    }
    public void doPost(HttpServletRequest request,HttpServletResponse
            response) throws ServletException,IOException{
        response.setContentType("text/html;charset=GBK");
        PrintWriter out=response.getWriter();
        request.setCharacterEncoding("GBK");
        String name=request.getParameter("user");
        【代码 1】                  //获得用户的会话对象
        session.setAttribute("myName", name);
        out.println("<htm><body>");
        out.println("您请求的 servlet 对象是:"+getServletName());
        out.println("<br>您的会话 ID 是:"+session.getId());
        out.println("<br>请单击请求另一个 servlet:");
        out.println("<br><a href=showMyName>请求另一个 servlet</a>");
        out.println("</body></htm>");
    }
    public void doGet(HttpServletRequest request,HttpServletResponse
            response) throws ServletException,IOException{
        doPost(request,response);
    }
}
```

ShowNameServlet.java

```java
package servlet;
import java.io.*;
import javax.servlet.*;
import javax.servlet.http.*;
public class ShowNameServlet extends HttpServlet {
    public void init(ServletConfig config) throws ServletException{
        super.init(config);
    }
    public void doPost(HttpServletRequest request,HttpServletResponse
            response) throws ServletException,IOException{
        response.setContentType("text/html;charset=GBK");
        PrintWriter out=response.getWriter();
        【代码2】                //获得用户的会话对象
        String name=(String)session.getAttribute("myName");
        out.println("<htm><body>");
        out.println("您请求的 servlet 对象是:"+getServletName());
        out.println("<br>您的会话 ID 是:"+session.getId());
        out.println("<br>您的会话中存储的用户名是:"+name);
        out.println("</body></htm>");
    }
    public void doGet(HttpServletRequest request,HttpServletResponse
            response) throws ServletException,IOException{
        doPost(request,response);
    }
}
```

3. 任务小结或知识扩展

用户的会话对象 session 可以在 JSP 页面中不作声明直接使用,而在 Servlet 类中必须先使用 request 对象获得用户的会话对象,然后再使用它。

4. 代码模板的参考答案

【代码 1】:HttpSession session=request.getSession(true);
【代码 2】:HttpSession session=request.getSession(true);

6.6.4 实践环节

请阐述在 JSP 页面中使用会话对象 session 和在 servlet 中使用会话对象 session 有什么不同?并举例说明。

6.7 小　　结

- Java Servlet 技术的核心思想就是在服务器端创建能响应用户请求的对象,该对象习惯上被称为一个 servlet 对象。
- 要想让 Web 服务器用 Servlet 类编译后的字节码文件创建 servlet 对象,必须为 Web 服务器编写一个部署文件 web.xml。

- servlet 对象第一次被请求加载时,服务器创建一个 servlet 对象,该对象调用 init 方法完成必要的初始化工作。当 servlet 对象成功创建后,servlet 对象就调用 service 方法来处理用户的请求并返回响应。
- 重定向的功能是将用户从当前 JSP 页面或 servlet 定向到另一个 JSP 页面或 servlet,以前的 request 中存放的变量全部失效,并进入一个新的 request 作用域;转发的功能是将用户对当前 JSP 页面或 servlet 的请求转发给另一个 JSP 页面或 servlet,以前的 request 中存放的变量不会失效。
- 一个用户在同一个 Web 服务的不同 servlet 中获取的 session 对象是完全相同的,但是不同用户的 session 对象互不相同。

习 题 6

1. servlet 对象是在服务器端还是在用户端被创建的?
2. 什么是转发?什么是重定向?它们有什么区别?
3. 简述 servlet 的生命周期与运行原理。
4. servlet 对象初始化时是调用 init 方法还是 service 方法?
5. servlet 对象是如何获得用户的会话对象的?
6. 假设创建 servlet 的类是 my.servlet.MyFirstServlet,创建的 servlet 对象的名字是 first,如果使用表单请求该 servlet 的话,表单的 action 的值是 isgo。我们应该如何为该 servlet 编写部署文件 web.xml?

基于 Servlet 的 MVC 模式

本章主要内容

- JSP 中的 MVC 模式
- 模型的生命周期与视图更新

我们已经学习了 JSP 和 Servlet 技术，使用它们可以开发出完整的 Web 应用程序。但有时需要把大量的 Java 代码写在 JSP 页面中，把 HTML 代码写在 Servlet 中，这样会造成代码编写不容易，日后维护也困难。因此，学习 Web 应用程序的设计模式是非常必要的。

本章将学习一种非常典型的 Web 应用程序的设计模式——基于 Servlet 的 MVC 模式。

7.1 JSP 中的 MVC 模式

7.1.1 核心知识

1. MVC 的概念

MVC 是 Model、View、Controller 的缩写，分别代表 Web 应用程序中的如下三种职责。

- 模型——用于存储数据以及处理用户请求的业务逻辑。
- 视图——向控制器提交数据，显示模型中的数据。
- 控制器——根据视图提出的请求，判断将请求和数据交给哪个模型处理，处理后的有关结果交给哪个视图更新显示。

2. JSP 中的 MVC 模式

JSP 中的 MVC 模式的具体实现如下。

- 模型：一个或多个 JavaBean 对象，用于存储数据（实体模型，由 JavaBean 类创建）和处理业务逻辑（业务模型，由一般的 Java 类创建）。
- 视图：一个或多个 JSP 页面，向控制器提交数据和为模型提供数据显示。JSP 页面主要使用 HTML 标记和 JavaBean 标记来显示数据。
- 控制器：一个或多个 servlet 对象，根据视图提交的请求进行控制，即把请求转发给处理业务逻辑的 JavaBean，并将处理结果存放到实体模型 JavaBean 中，输出给视图显示。

JSP 中的 MVC 模式的流程如图 7.1 所示。

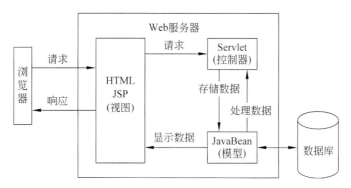

图 7.1　JSP 中的 MVC 模式

7.1.2　能力目标

理解 JSP 中的 MVC 模式的实现过程。

7.1.3　任务驱动

1. 任务的主要内容

编写一个简单的 Web 应用程序：用户登录验证程序。视图 login.jsp 提交数据请求（用户名和密码）；控制器 controllerServlet（ControllerServlet 类负责创建）接收请求信息，然后把请求信息封装在 user（UserBean 类负责创建）实体模型中，控制器把 user 模型传递给 userCheck 业务模型去处理（UserCheckBean 类负责创建）；如果用户名和密码输入正确，返回 success.jsp 页面，否则返回 login.jsp 页面。程序运行效果如图 7.2 所示。

图 7.2　登录验证程序

2. 任务的代码模板

将 ControllerServlet.java 中的【代码】替换为 Java 代码。

web.xml

```
<?xml version="1.0" encoding="UTF-8"?>
<web-app>
<servlet>
    <servlet-name>controllerServlet</servlet-name>
    <servlet-class>servlet.ControllerServlet</servlet-class>
</servlet>
<servlet-mapping>
    <servlet-name>controllerServlet</servlet-name>
    <url-pattern>/isLogin</url-pattern>
</servlet-mapping>
</web-app>
```

UserBean.java（实体模型）

```java
package entity.bean;
public class UserBean {
    String name;
    String pwd;
    public UserBean(){
    }
    public String getName() {
        return name;
    }
    public void setName(String name) {
        this.name=name;
    }
    public String getPwd() {
        return pwd;
    }
    public void setPwd(String pwd) {
        this.pwd=pwd;
    }
}
```

UserCheckBean.java（业务模型）

```java
package busynees.bean;
import entity.bean.UserBean;
public class UserCheckBean {
    //验证登录
    public boolean validate(UserBean user){
        if(user!=null&&user.getName().equals("JSPMVC")){
            if(user.getPwd().equals("MVC")){
                return true;
            }
            return false;
        }
        return false;
    }
}
```

login.jsp（视图）

```jsp
<%@page language="java" contentType="text/html; charset=GBK" pageEncoding="GBK"%>
<html>
    <head>
        <title>login.jsp</title>
    </head>
    <body>
        <form action="isLogin" method="post">
        <table>
            <tr>
                <td>用户名:</td>
```

```
            <td><input type="text" name="name"/></td>
        </tr>
        <tr>
            <td>密　　码:</td>
            <td><input type="password" name="pwd"/></td>
        </tr>
        <tr>
            <td><input type="submit" value="提交"/></td>
            <td><input type="reset" value="重置"/></td>
        </tr>
    </table>
    </form>
</body>
</html>
```

success.jsp（视图）

```
<%@page language="java" contentType="text/html; charset=GBK" pageEncoding="GBK"%>
<%@page import="entity.bean.UserBean" %>
<html>
<head>
<title>success.jsp</title>
</head>
<body>
<jsp:useBean id="userBean" type="entity.bean.UserBean" scope="request"/>
恭喜<jsp:getProperty property="name" name="userBean"/>登录成功！
</body>
</html>
```

ControllerServlet.java（控制器）

```
package servlet;
import java.io.IOException;
import java.io.PrintWriter;
import javax.servlet.RequestDispatcher;
import javax.servlet.ServletConfig;
import javax.servlet.ServletException;
import javax.servlet.http.HttpServlet;
import javax.servlet.http.HttpServletRequest;
import javax.servlet.http.HttpServletResponse;
import busynees.bean.UserCheckBean;
import entity.bean.UserBean;
public class ControllerServlet extends HttpServlet {
    public void init(ServletConfig config) throws ServletException{
        super.init(config);
    }
    public void doPost(HttpServletRequest request,HttpServletResponse response)
            throws IOException,ServletException{
        request.setCharacterEncoding("GBK");          //设置编码,防止中文乱码
        String name=【代码 1】                          //获取客户提交的信息
        String password=【代码 2】                      //获取客户提交的信息
```

```
    【代码 3】                                    //实例化实体模型 user
    user.setName(name);                          //把数据存在模型 user 中
    user.setPwd(password);                       //把数据存在模型 user 中
    【代码 4】                                    //实例化业务模型 userCheck
    if(userCheck.validate(user)){
        request.setAttribute("userBean", user);  //把装有数据的实体模型 user
        RequestDispatcher dis=request.getRequestDispatcher("success.jsp");
        dis.forward(request, response);
    }else{
        response.sendRedirect("login.jsp");
    }
}
public void doGet(HttpServletRequest request,HttpServletResponse response)
        throws IOException,ServletException{
    doPost(request,response);
}
}
```

3. 任务小结或知识扩展

在 JSP 的 MVC 模式中，控制器 servlet 创建的实体模型 JavaBean 也涉及生命周期，生命周期分别为 request、session 和 application。下面以任务中的实体模型 user 来讨论这 3 种生命周期的模型的用法。

1) request 周期的模型

使用 request 周期的模型一般需要以下几个环节。

（1）创建模型并把数据保存到模型中

在 servlet 中需要这样的代码。

```
UserBean user=new UserBean();           //实例化模型 user
user.setName(name);                     //把数据存入模型 user 中
user.setPwd(password);                  //把数据存入模型 user 中
```

（2）将模型保存到 request 对象中并转发给视图 JSP

在 servlet 中需要这样的代码。

```
request.setAttribute("userBean", user);
                            //把装有数据的模型 user 输出给视图 success.jsp 页面
RequestDispatcher dis=request.getRequestDispatcher("success.jsp");
dis.forward(request, response);
```

request.setAttribute("userBean",user)这句代码指定了查找 JavaBean 的关键字，并决定了 JavaBean 的生命周期为 request。

（3）视图更新

servlet 所转发的页面，比如 success.jsp 页面，必须使用 useBean 标记获得 servlet 所创建的 JavaBean 对象（视图不负责创建 JavaBean）。在 JSP 页面需要使用这样的代码。

```
<jsp:useBean id="userBean" type="entity.bean.UserBean" scope="request"/>
<jsp:getProperty property="name" name="userBean"/>
```

标记中的 id 就是 servlet 所创建的模型 JavaBean，它和 request 对象中的关键字对应。因为在视图中不创建 JavaBean 对象，所以在 useBean 标记中使用 type 属性，而不使用 class 属性。useBean 标记中的 scope 必须和存储模型时的范围（request）一致。

2) session 周期的模型

使用 session 周期的模型一般需要以下几个环节。

(1) 创建模型并把数据保存到模型中

在 servlet 中需要这样的代码。

```
UserBean user=new UserBean();           //实例化模型 user
user.setName(name);                      //把数据存在模型 user 中
user.setPwd(password);                   //把数据存在模型 user 中
```

(2) 将模型保存到 session 对象中并转发给视图 JSP

在 servlet 中需要这样的代码。

```
session.setAttribute("userBean", user);
                         //把装有数据的模型 user 输出给视图 success.jsp 页面
RequestDispatcher dis=request.getRequestDispatcher("success.jsp");
dis.forward(request, response);
```

session.setAttribute("userBean",user)这句代码指定了查找 JavaBean 的关键字，并决定了 JavaBean 的生命周期为 session。

(3) 视图更新

servlet 所转发的页面，比如 success.jsp 页面，必须使用 useBean 标记获得 servlet 所创建的 JavaBean 对象（视图不负责创建 JavaBean）。在 JSP 页面需要使用这样的代码。

```
<jsp:useBean id="userBean" type="entity.bean.UserBean" scope="session"/>
<jsp:getProperty property="name" name="userBean"/>
```

标记中的 id 就是 servlet 所创建的模型 JavaBean，它和 session 对象中的关键字对应。因为在视图中不创建 JavaBean 对象，所以在 useBean 标记中使用 type 属性，而不使用 class 属性。useBean 标记中的 scope 必须和存储模型时的范围（session）一致。

注：对于生命周期为 session 的模型，servlet 不仅可以使用 RequestDispatcher 对象转发给 JSP 页面，也可以使用 response 的重定向方法（sendRedirect）定向到 JSP 页面。

3) application 周期的模型

使用 application 周期的模型一般需要以下几个环节。

(1) 创建模型并把数据保存到模型中

在 servlet 中需要这样的代码。

```
UserBean user=new UserBean();           //实例化模型 user
user.setName(name);                      //把数据存在模型 user 中
user.setPwd(password);                   //把数据存在模型 user 中
```

(2) 将模型保存到 application 对象中并转发给视图 JSP

在 servlet 中需要这样的代码。

```
application.setAttribute("userBean", user);
                        //把装有数据的模型 user 输出给视图 success.jsp 页面
RequestDispatcher dis=request.getRequestDispatcher("success.jsp");
dis.forward(request, response);
```

application.setAttribute("userBean"，user)这句代码指定了查找 JavaBean 的关键字，并决定了 JavaBean 的生命周期为 application。

（3）视图更新

servlet 所转发的页面，比如 success.jsp 页面，必须使用 useBean 标记获得 servlet 所创建的 JavaBean 对象（视图不负责创建 JavaBean）。在 JSP 页面需要使用这样的代码。

```
<jsp:useBean id="userBean" type="entity.bean.UserBean" scope="application"/>
<jsp:getProperty property="name" name="userBean"/>
```

标记中的 id 就是 servlet 所创建的模型 JavaBean，它和 application 对象中的关键字对应。因为在视图中不创建 JavaBean 对象，所以在 useBean 标记中使用 type 属性，而不使用 class 属性。useBean 标记中的 scope 必须和存储模型时的范围（application）一致。

注：对于生命周期为 application 的模型，servlet 不仅可以使用 RequestDispatcher 对象转发给 JSP 页面，也可以使用 response 的重定向方法（sendRedirect）定向到 JSP 页面。

4. 代码模板的参考答案

【代码 1】：`request.getParameter("name");`
【代码 2】：`request.getParameter("pwd");`
【代码 3】：`UserBean user=new UserBean();`
【代码 4】：`UserCheckBean userCheck=new UserCheckBean();`

7.1.4 实践环节

把任务中实体模型 user 的生命周期改为 session，并运行程序。

7.2 使用 MVC 模式查询数据库

7.2.1 核心知识

使用 MVC 模式设计 Web 应用时，尽量把实体模型与业务模型分开实现，方便以后维护。例如，在"使用 MVC 模式查询数据库"这个 Web 应用中，数据的封装由实体模型 Goods 完成，处理数据由业务模型 GoodsDao 完成，与数据库连接、关闭等操作由业务模型 DataBaseBean 完成。

7.2.2 能力目标

灵活使用 MVC 模式设计 Web 应用。

7.2.3 任务驱动

1. 任务的主要内容

设计一个 Web 应用，有两个 JSP 页面（addGoods.jsp 和 showAllGoods.jsp）、3 个 JavaBean（实体模型 Goods、业务模型 GoodsDao 和业务模型 DataBaseBean）和一个 servlet（GoodsControllerServlet）。用户在 JSP 页面 addGoods.jsp 中输入商品的信息，提交给 servlet，该 servlet 负责添加商品（调用业务模型 GoodsDao 的 addGoods 方法）、查询商品（调用业务模型 GoodsDao 的 getAllGoods 方法），并把查询结果显示在 showAllGoods.jsp 页面中。Web 应用中使用了 5.2 节曾经使用过的数据表 goodsInfo。

用户需要为 web.xml 文件添加如下子标记。

```
<servlet>
    <servlet-name>addServlet</servlet-name>
    <servlet-class>servlet.GoodsControllerServlet</servlet-class>
</servlet>
<servlet-mapping>
    <servlet-name>addServlet</servlet-name>
    <url-pattern>/addServlet</url-pattern>
</servlet-mapping>
```

页面运行效果如图 7.3 与图 7.4 所示。

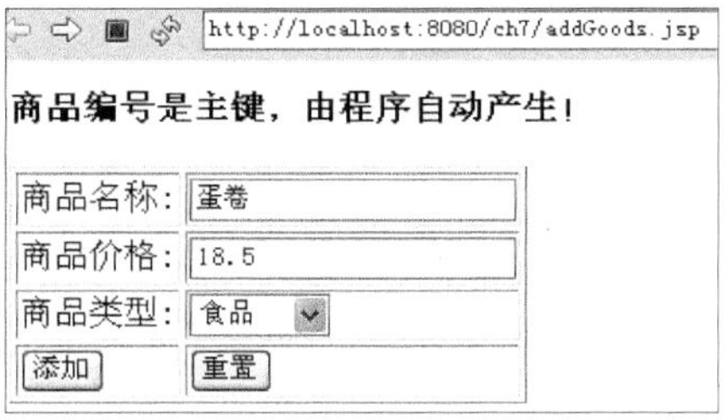

图 7.3　商品信息输入页面

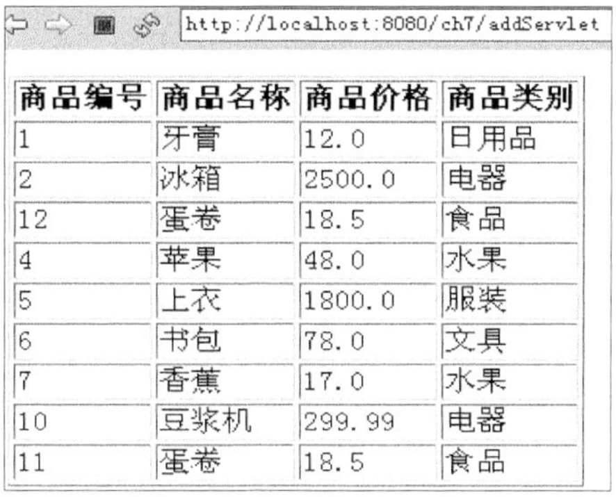

商品编号	商品名称	商品价格	商品类别
1	牙膏	12.0	日用品
2	冰箱	2500.0	电器
12	蛋卷	18.5	食品
4	苹果	48.0	水果
5	上衣	1800.0	服装
6	书包	78.0	文具
7	香蕉	17.0	水果
10	豆浆机	299.99	电器
11	蛋卷	18.5	食品

图 7.4　商品信息显示页面

2. 任务的代码模板

将 GoodsControllerServlet.java 中的【代码】替换为 Java 代码。

Goods.java（实体模型）

```
package entity.bean;
public class Goods {
    int goodsId;
    String goodsName;
    double goodsPrice;
    String goodsType;
    StringBuffer result;
    public Goods(){
```

```java
    }
    public int getGoodsId() {
        return goodsId;
    }
    public void setGoodsId(int goodsId) {
        this.goodsId=goodsId;
    }
    public String getGoodsName() {
        return goodsName;
    }
    public void setGoodsName(String goodsName) {
        this.goodsName=goodsName;
    }
    public double getGoodsPrice() {
        return goodsPrice;
    }
    public void setGoodsPrice(double goodsPrice) {
        this.goodsPrice=goodsPrice;
    }
    public String getGoodsType() {
        return goodsType;
    }
    public void setGoodsType(String goodsType) {
        this.goodsType=goodsType;
    }
    public StringBuffer getResult() {
        return result;
    }
    public void setResult(StringBuffer result) {
        this.result=result;
    }
}
```

GoodsDao.java（业务模型）

```java
package busynees.bean;
import java.sql.*;
import java.util.ArrayList;
import entity.bean.Goods;
public class GoodsDao {
    //获得新添加的商品编号
    public synchronized int getID(){
        Connection con=DataBaseBean.getCon();
        PreparedStatement ps=null;
        ResultSet rs=null;
        int id=0;
        try {
            ps=con.prepareStatement("select max(goodsId) from goodsInfo");
            rs=ps.executeQuery();
            if(rs.next()){
                id=rs.getInt(1)+1;
```

```java
                }
            } catch (SQLException e) {
                //TODO Auto-generated catch block
                e.printStackTrace();
            }finally{
                DataBaseBean.close(rs);
                DataBaseBean.close(ps);
                DataBaseBean.close(con);
            }
            return id;
    }
    //添加商品
    public boolean addGoods(Goods goods){
        Connection con=DataBaseBean.getCon();
        PreparedStatement ps=null;
        try {
            ps=con.prepareStatement("insert into goodsInfo values(?,?,?,?)");
            ps.setInt(1, goods.getGoodsId());
            ps.setString(2, goods.getGoodsName());
            ps.setDouble(3, goods.getGoodsPrice());
            ps.setString(4, goods.getGoodsType());
            int i=ps.executeUpdate();
            if(i>0)
                return true;
        } catch (SQLException e) {
            //TODO Auto-generated catch block
            e.printStackTrace();
        }finally{
            DataBaseBean.close(ps);
            DataBaseBean.close(con);
        }
        return false;
    }
    //查询商品
    public StringBuffer getAllGoods(){
        Connection con=DataBaseBean.getCon();
        PreparedStatement ps=null;
        ResultSet rs=null;
        StringBuffer str=new StringBuffer();
        try {
            ps=con.prepareStatement("select * from goodsInfo");
            rs=ps.executeQuery();
            str.append("<table border=1>" +
                    "<tr>" +
                    "<th>商品编号</th>" +
                    "<th>商品名称</th>" +
                    "<th>商品价格</th>" +
                    "<th>商品类别</th>" +
                    "</tr>");
            while(rs.next()){
                str.append("<tr>");
```

```
                str.append("<td>"+rs.getString(1)+"</td>");
                str.append("<td>"+rs.getString(2)+"</td>");
                str.append("<td>"+rs.getString(3)+"</td>");
                str.append("<td>"+rs.getString(4)+"</td>");
                str.append("</tr>");
            }
            str.append("</table>");
        } catch (SQLException e) {
            //TODO Auto-generated catch block
            e.printStackTrace();
        }finally{
            DataBaseBean.close(rs);
            DataBaseBean.close(ps);
            DataBaseBean.close(con);
        }
        return str;
    }
}
```

DataBaseBean.java（业务模型）

```
package busynees.bean;
import java.sql.Connection;
import java.sql.DriverManager;
import java.sql.PreparedStatement;
import java.sql.ResultSet;
import java.sql.SQLException;
public class DataBaseBean {
    public static Connection getCon(){
        Connection con=null;
        try {
            Class.forName("oracle.jdbc.driver.OracleDriver");
        } catch (ClassNotFoundException e) {
            e.printStackTrace();
        }
        try {
        con=DriverManager.getConnection("jdbc:oracle:thin:@127.0.0.1:1521:orcl","system","system");
        } catch (SQLException e) {
            e.printStackTrace();
        }
        return con;
    }
    public static void close(ResultSet rs){
        if(rs!=null){
            try {
                rs.close();
            } catch (SQLException e) {
                e.printStackTrace();
            }
        }
```

```java
    }
    public static void close(PreparedStatement ps){
        if(ps!=null){
            try {
                ps.close();
            } catch (SQLException e) {
                e.printStackTrace();
            }
        }
    }
    public static void close(Connection con){
        if(con!=null){
            try {
                con.close();
            } catch (SQLException e) {
                e.printStackTrace();
            }
        }
    }
}
```

addGoods.jsp(视图1)

```jsp
<%@page language="java" contentType="text/html; charset=GBK" pageEncoding="GBK"%>
<html>
<head>
<title>addGoods.jsp</title>
</head>
<body>
    <h4>商品编号是主键,由程序自动产生!</h4>
    <form action="addServlet" method="post">
    <table border="1">
        <tr>
            <td>商品名称:</td>
            <td><input type="text" name="goodsName"/></td>
        </tr>

        <tr>
            <td>商品价格:</td>
            <td><input type="text" name="goodsPrice"/></td>
        </tr>

        <tr>
            <td>商品类型:</td>
            <td>
                <select name="goodsType">
                    <option value="日用品">日用品
                    <option value="电器">电器
                    <option value="食品">食品
                    <option value="水果">水果
```

```
                <option value="服装">服装
                <option value="文具">文具
                <option value="其他">其他
            </select>
        </td>
    </tr>

    <tr>
        <td><input type="submit" value="添加"></td>
        <td><input type="reset" value="重置"></td>
    </tr>
    </table>
    </form>
</body>
</html>
```

showAllGoods.jsp（视图2）

```
<%@ page language="java" contentType="text/html; charset=GBK" pageEncoding="GBK"%>
<%@ page import="entity.bean.Goods" %>
<html>
<head>
<title>showAllGoods.jsp</title>
</head>
<body>
<jsp:useBean id="goods" type="entity.bean.Goods" scope="request"></jsp:useBean>
<jsp:getProperty property="result" name="goods"/>
</body>
</html>
```

GoodsControllerServlet.java（控制器）

```
package servlet;
import java.io.IOException;
import javax.servlet.RequestDispatcher;
import javax.servlet.ServletConfig;
import javax.servlet.ServletException;
import javax.servlet.http.HttpServlet;
import javax.servlet.http.HttpServletRequest;
import javax.servlet.http.HttpServletResponse;
import busynees.bean.GoodsDao;
import entity.bean.Goods;
public class GoodsControllerServlet extends HttpServlet {
    public void init(ServletConfig config) throws ServletException{
        super.init(config);
    }
    public void doPost ( HttpServletRequest request, HttpServletResponse
        response)
        throws IOException,ServletException{
        request.setCharacterEncoding("GBK");           //设置编码,防止中文乱码
```

```
        String goodsName=request.getParameter("goodsName");
                                              //获得视图提交的信息
        String goodsPrice=request.getParameter("goodsPrice");
        String goodsType=request.getParameter("goodsType");
        Goods goods=【代码 1】                     //创建实体模型 goods
        GoodsDao gd=【代码 2】                     //创建业务模型 gd
        goods.setGoodsId(gd.getID());            //把数据存储到实体模型中
        goods.setGoodsName(goodsName);           //把数据存储到实体模型中
        goods.setGoodsPrice(Double.parseDouble(goodsPrice));
                                                 //把数据存储到实体模型中
        goods.setGoodsType(goodsType);           //把数据存储到实体模型中
        if(gd.addGoods(goods)){
            //调用 getAllGoods()获得所有商品信息,添加到实体模型中
            goods.setResult(gd.getAllGoods());
            request.setAttribute("goods", goods);  //把实体模型保存到 request 里面
            RequestDispatcher dis=request.getRequestDispatcher("showAllGoods.
            jsp");
            dis.forward(request, response);
        }else{
            response.sendRedirect("addGoods.jsp");
        }
    }
    public void doGet(HttpServletRequest request,HttpServletResponse response)
            throws IOException,ServletException{
        doPost(request,response);
    }
}
```

3. 任务小结或知识扩展

使用 MVC 模式设计 Web 应用时,控制器尽量不处理数据,它只是起到控制转发的作用,而数据交给业务模型处理。如,任务中的控制器。

4. 代码模板的参考答案

【代码 1】:`new Goods();`
【代码 2】:`new GoodsDao();`

7.2.4 实践环节

使用 MVC 模式设计一个 Web 应用,用户通过 JSP 页面 inputNumber.jsp 输入两个操作数,并选择一种运算符,单击"提交"按钮后,调用 HandleComputer.java 这个 Servlet。在 HandleComputer.java 中获取用户输入的数字和运算符并将这些内容放入 ComputerBean.java 这个 JavaBean 中,在 showResult.jsp 中调用 JavaBean 显示计算的结果。

7.3 小　　结

MVC 模式将"视图"、"模型"和"控制器"有效地组合。在 JSP 页面的 MVC 模式中,视图是一个或多个 JSP 页面,作用是向控制器提交数据和为模型提供数据显示;模型是一个

或多个 JavaBean 对象,用于存储数据(实体模型,由 JavaBean 类创建)和处理业务逻辑(业务模型,由一般的 Java 类创建);控制器是一个或多个 servlet 对象,根据视图提交的请求进行控制,即把请求转发给处理业务逻辑的 JavaBean,并将处理结果存放到实体模型 JavaBean 中,输出给视图显示。

习 题 7

1. 以下不属于 MVC 设计模式中 3 个模块的是(　　)。
   A. 模型　　　　　B. 表示层　　　　C. 视图　　　　　D. 控制器
2. 在 MVC 模式中,(　　)用于客户端应用程序的图形数据表示,与实际数据处理无关。
   A. 模型　　　　　B. 视图　　　　　C. 控制器　　　　D. 数据
3. 在 MVC 设计模式中,(　　)接收用户请求数据。
   A. HTML　　　　B. JSP　　　　　C. Servlet　　　　D. 业务类
4. 使用 MVC 模式设计 Web 应用有什么好处?
5. MVC 中的模型是由 servlet 负责创建,还是由 JSP 页面负责创建?

过滤器

本章主要内容

- 过滤器的概念
- 过滤器的运行原理
- 过滤器的实际应用

在开发一个网站时,可能有这样的需求:某些页面只希望几个特定的用户浏览。对于这样的访问权限的控制,该如何实现呢?过滤器(Filter)就可以实现上述需求。过滤器位于服务器(Servlet)处理请求之前或服务器响应请求之前。也就是说,它可以过滤浏览器对服务器的请求,也可以过滤服务器对浏览器的响应,如图 8.1 所示。

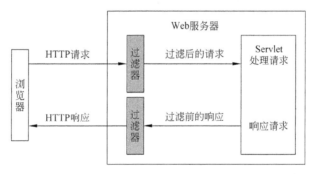

图 8.1 过滤器

如何编写过滤器类,又如何使用过滤器类,这些内容将在本章作重点介绍。

8.1 Filter 类与 filter 对象

8.1.1 核心知识

编写一个过滤器类很简单,只要实现 javax.servlet 包中的 Filter 接口。实现 Filter 接口的类习惯上称为一个 Filter 类,这样的类创建的对象习惯上又称为 filter 对象。

8.1.2 能力目标

理解 Filter 类与 filter 对象的概念。

8.1.3 任务驱动

1. 任务的主要内容

新建一个 Web 工程 ch8，在该 Web 工程中编写一个简单的 Filter 类 MyFirstFilter，Filter 类实现如下功能：

不管用户请求该 Web 工程的哪个页面或 servlet，都会在浏览器中出现"首先执行过滤器"这样的响应信息。

2. 任务的代码模板

将下列 MyFirstFilter.java 中的【代码】替换为 Java 代码。

MyFirstFilter.java

```java
package filters;
import java.io.IOException;
import java.io.PrintWriter;
import javax.servlet.Filter;
import javax.servlet.FilterChain;
import javax.servlet.FilterConfig;
import javax.servlet.ServletException;
import javax.servlet.ServletRequest;
import javax.servlet.ServletResponse;
public class MyFirstFilter implements 【代码】{
    public void destroy() {
    }
    public void doFilter(ServletRequest request,
            ServletResponse response,
            FilterChain chain) throws IOException, ServletException {
        //设置响应类型
        response.setContentType("text/html;charset=GBK");
        //获得输出对象 out
        PrintWriter out=response.getWriter();
        //在浏览器中输出
        out.print("首先执行过滤器<br>");
        //执行下一个过滤器
        chain.doFilter(request, response);
    }
    public void init(FilterConfig fConfig) throws ServletException {
    }
}
```

3. 任务小结或知识扩展

从任务中 MyFirstFilter 类的源代码可以看出：Filter 接口与 Servlet 接口很类似，同样都有 init() 与 destroy() 方法，还有一个 doFilter() 方法类似于 Servlet 接口的 service() 方法。下面分别介绍这 3 种方法的功能。

(1) public void init(FilterConfig fConfig) throws ServletException

方法的功能是初始化过滤器对象。方法中的参数 fConfig 是 FilterConfig 的对象，该对象代表 web.xml 中为过滤器定义的对象。如果在 web.xml 中为过滤器设置了初始参数，则可以通过 FilterConfig 的 getInitParameter(String paramName)方法获得初始参数值。

（2）public void doFilter（ServletRequest request，ServletResponse response，FilterChain chain）throws IOException，ServletException

当 Web 服务器使用 servlet 对象调用 service()方法处理请求前，发现在应用某个过滤器时，Web 服务器就会自动调用该过滤器的 doFilter()方法。在 doFilter()方法中就会有这样一条语句。

```
chain.doFilter(request, response);
```

如果执行了该语句，就会执行下一个过滤器（根据＜filter-mapping＞在 web.xml 中出现的先后顺序执行过滤器），如果没有下一个过滤器，就返回请求目标程序。如果因为某个原因没有执行"chain.doFilter(request，response)；"，则请求就不会继续交给以后的过滤器或请求目标程序，这时就是所谓的拦截请求。

（3）public void destroy()

当 Web 服务器终止服务时，destroy 方法会被执行，使 filter 对象消亡。

编写完 Filter 类的源文件，并编译了源文件，这时 Web 服务器是不是就可以运行 filter 对象呢？不可以。需要部署 filter 之后，Web 服务器才可以运行 filter 对象。

4. 代码模板的参考答案

【代码】：Filter

8.1.4 实践环节

尝试寻找任务中的 Filter 类编译后的字节码文件。

8.2 filter 对象的部署与运行

8.2.1 核心知识

要想让 Web 服务器用 Filter 类编译后的字节码文件创建 filter 对象，必须在 Web 工程的 web.xml 文件里部署自己的 filter。

8.2.2 能力目标

掌握部署过滤器的方法。

8.2.3 任务驱动

1. 任务的主要内容

- 部署 filter；
- 运行 filter。

2. 任务模板

请按下列步骤进行操作。

(1) 部署 filter

为了在 web.xml 文件里部署 8.1 节中的 MyFirstFilter,需要在 web.xml 文件里找到 <web-app></web-app>标记,然后在<web-app></web-app>标记中添加如下内容。

```
<filter>
    <filter-name>myFirstFilter</filter-name>
    <filter-class>filters.MyFirstFilter</filter-class>
</filter>
<filter-mapping>
    <filter-name>myFirstFilter</filter-name>
    <url-pattern>/*</url-pattern>
</filter-mapping>
```

(2) 运行 filter

只要用户请求的 URL 和<filter-mapping>的子标记<url-pattern>指定的模式匹配的话,Web 服务器就会自动调用该 filter 的 doFilter()方法。如:8.1 节中的 MyFirstFilter 过滤器在 web.xml 中的<url-pattern>指定值为/*,"/*"代表任何页面或 servlet 的请求。

3. 任务小结或知识扩展

(1) <filter>标记及其子标记

web.xml 文件中可以有若干个<filter>标记,该标记的内容由 Web 服务器负责处理。<filter>标记中有两个子标记:<filter-name>和<filter-class>,其中<filter-name>子标记的内容是 Web 服务器创建的 filter 对象的名字。web.xml 文件中可以有若干个<filtert>标记,但要求它们的<filter-name>子标记的内容互不相同。<filter-class>子标记的内容是指定 Web 服务器用哪个类来创建 filter 对象,如果 filter 对象已经创建,那么 Web 服务器就不再使用指定的类创建。

如果在过滤器初始化时,需要读取一些参数的值,则可以在<filter>标记中使用<init-param>子标记设置。例如:

```
<filter>
    <filter-name>myFirstFilter</filter-name>
    <filter-class>filters.MyFirstFilter</filter-class>
    <init-param>
        <param-name>encoding</param-name>
        <param-value>GBK</param-value>
    </init-param>
</filter>
```

那么就可以在 filter 的 init()方法中,使用参数 fConfig(FilterConfig 的对象)调用 FilterConfig 的 getInitParameter(String paramName)方法获得参数值。例如:

```
public void init(FilterConfig fConfig) throws ServletException{
    String en=fConfig.getInitParameter("encoding");
}
```

(2) <filter-mapping>标记及其子标记

web.xml 文件中出现一个<filter>标记就会对应地出现一个<filter-mapping>标记。

＜filter-mapping＞标记中也有两个子标记：＜filter-name＞和＜url-pattern＞。其中＜filter-name＞子标记的内容是 Web 服务器创建的 filter 对象的名字（该名字必须和＜filter＞标记的子标记＜filter-name＞的内容相同）；＜url-pattern＞子标记用来指定用户用怎样的模式请求 filter 对象。如果某个 URL 或 servlet 需应用多个过滤器，则根据＜filter-mapping＞标记在 web.xml 中出现的先后顺序执行过滤器。

4. 代码模板的参考答案

无参考答案

8.2.4 实践环节

按照本节的任务内容将 8.1 节中的过滤器 MyFirstFilter 部署成功，并运行 Web 应用程序测试该过滤器。

8.3 登录验证过滤器的实现

8.3.1 核心知识

在 Web 工程中，某些页面或 servlet 只有用户登录成功才能访问。如果直接在应用程序每个相关的源代码中进行判断用户是否登录成功，并不是科学的做法。我们可以实现一个登录验证过滤器，在 Web 工程的 web.xml 中设置并使用该过滤器，就可以不用在每个相关的源代码中验证用户是否登录成功。

8.3.2 能力目标

理解登录验证过滤器的实现方法。

8.3.3 任务驱动

1. 任务的主要内容

新建一个 Web 工程 loginValidate，在该 Web 工程中至少编写两个 JSP 页面：login.jsp 与 loginSuccess.jsp，一个 servlet（由 LoginServlet.java 负责创建）。用户在 login.jsp 页面中输入用户名和密码后，提交给 servlet，在 servlet 中判断用户名和密码是否正确，若正确跳转到 loginSuccess.jsp，若错误回到 login.jsp 页面。但该 Web 工程有另外一个要求：除了访问 login.jsp 页面外，别的页面或 servlet 都不能直接访问，必须先登录成功才能访问。我们在设计这个 Web 工程时，编写了一个登录验证过滤器并在该 Web 工程中使用。

页面运行效果如图 8.2(a)～图 8.2(c)所示。

2. 任务的代码模板

将下列 LoginFilter.java 中的【代码】替换为 Java 代码。

web.xml

```
<web-app>
    <filter>
```

```xml
        <filter-name>loginValidateFilter</filter-name>
        <filter-class>filters.LoginFilter</filter-class>
        <init-param>
            <param-name>login_uri</param-name>
            <param-value>/login.jsp</param-value>
        </init-param>
        <init-param>
            <param-name>login_Servlet</param-name>
            <param-value>/isLogin</param-value>
        </init-param>
    </filter>
    <filter-mapping>
        <filter-name>loginValidateFilter</filter-name>
        <url-pattern>/*</url-pattern>
    </filter-mapping>

    <servlet>
        <servlet-name>loginServlet</servlet-name>
        <servlet-class>servlet.LoginServlet</servlet-class>
    </servlet>
    <servlet-mapping>
        <servlet-name>loginServlet</servlet-name>
        <url-pattern>/isLogin</url-pattern>
    </servlet-mapping>
</web-app>
```

(a) 登录画面　　　　　　　　　　(b) 没有登录成功直接运行loginSuccess.jsp

(c) 登录成功页面

图 8.2　8.3 节任务的页面运行效果图

LoginFilter.java（过滤器）

```java
package filters;
import java.io.IOException;
import java.io.PrintWriter;
import javax.servlet.Filter;
import javax.servlet.FilterChain;
import javax.servlet.FilterConfig;
import javax.servlet.ServletException;
import javax.servlet.ServletRequest;
import javax.servlet.ServletResponse;
import javax.servlet.http.HttpServletRequest;
```

```java
import javax.servlet.http.HttpServletResponse;
import javax.servlet.http.HttpSession;
public class LoginFilter implements Filter {
    private String logon_page;                //登录页面
    private String logon_servlet;             //登录 servlet 请求
    //消灭 filter 方法
    public void destroy() {
    }
    //过滤器服务方法
    public void doFilter(ServletRequest request, ServletResponse response,
            FilterChain chain) throws IOException, ServletException {
        HttpServletRequest req=(HttpServletRequest) request;
        HttpServletResponse resp=(HttpServletResponse) response;
        resp.setContentType("text/html;");
        resp.setCharacterEncoding("GBK");
        HttpSession session=req.getSession();
        PrintWriter out=resp.getWriter();
        //得到用户请求的 URI
        String request_uri=req.getRequestURI();
        //得到 web 应用程序的上下文路径
        String ctxPath=req.getContextPath();
        //去除上下文路径,得到剩余部分的路径
        String uri=request_uri.substring(ctxPath.length());
        //判断用户访问的是否是登录页面或提交登录请求
        if (uri.equals(logon_page)||uri.equals(logon_servlet)) {
            //执行下一个过滤器
            chain.doFilter(request, response);
        } else {
            //如果访问的不是登录页面,则判断用户是否已经登录
            if (null !=session.getAttribute("user")
                    && "" !=session.getAttribute("user")) {
                //执行下一个过滤器
                chain.doFilter(request, response);
            } else {
                out.println("您没有登录,请先登录!3 秒钟后回到登录页面。");
                resp.setHeader("refresh", "3;url=" +ctxPath +logon_page);
                return;
            }
        }
    }
    //过滤器初始化方法
    public void init(FilterConfig config) throws ServletException {
        //从 web.xml 的部署描述符中获取登录页面
        logon_page=【代码 1】                  //获得参数 login_uri 的值
        logon_servlet =【代码 2】              //获得参数 login_uri 的值
    }
}
```

LoginServlet.java

```java
package servlet;
import java.io.IOException;
import javax.servlet.ServletException;
import javax.servlet.http.HttpServlet;
import javax.servlet.http.HttpServletRequest;
import javax.servlet.http.HttpServletResponse;
import javax.servlet.http.HttpSession;
public class LoginServlet extends HttpServlet {
    protected void doGet(HttpServletRequest request,
            HttpServletResponse response) throws ServletException, IOException {
        String username=request.getParameter("name");
        String password=request.getParameter("pwd");
        if(username!=null&&username.equals("filter")){
            if(password!=null&&password.equals("filter")){
                HttpSession session=request.getSession();
                session.setAttribute("user", username);
                response.sendRedirect("loginSuccess.jsp");
            }else{
                response.sendRedirect("login.jsp");
            }
        }else{
            response.sendRedirect("login.jsp");
        }
    }
    protected void doPost(HttpServletRequest request,
            HttpServletResponse response) throws ServletException, IOException {
        doGet(request,response);
    }
}
```

login.jsp

```jsp
<%@ page language="java" contentType="text/html; charset=GBK" pageEncoding="GBK"%>
<html>
    <head>
        <title>login.jsp</title>
    </head>
    <body bgcolor="lightPink">
        <form action="isLogin" method="post">
            <table>
                <tr>
                    <td>用户名:</td>
                    <td><input type="text" name="name"/></td>
                </tr>
                <tr>
                    <td>密  码:</td>
                    <td><input type="password" name="pwd"/></td>
                </tr>
```

```
            <tr>
                <td><input type="submit" value="提交"/></td>
                <td><input type="reset" value="重置"/></td>
            </tr>
        </table>
    </form>
</body>
</html>
```

loginSuccess.jsp

```
<%@page language="java" contentType="text/html; charset=GBK" pageEncoding="GBK"%>
<html>
<head>
<title>loginSuccess.jsp</title>
</head>
<body>
    <%
    String username=(String)session.getAttribute("user");
    %>
    恭喜<%=username %>登录成功!
</body>
</html>
```

3. 任务小结或知识扩展

任务中的过滤器,要首先检查用户请求的 URL 是不是 login.jsp 或者登录请求(isLogin),这两个值都放在了过滤器的初始化参数中。如果用户访问的是 login.jsp 或者登录请求的话,过滤器就执行 chain..doFilter()继续请求。如果用户访问的不是 login.jsp 或者登录请求的话,过滤器先判断用户是否登录成功,若登录成功,执行 chain..doFilter()继续请求,否则重定向到 login.jsp。

4. 代码模板的参考答案

【代码 1】: config.getInitParameter("login_uri");
【代码 2】: config.getInitParameter("login_Servlet");

8.3.4 实践环节

在任务的 Web 工程 loginValidate 中再新建几个 JSP 页面,在没有登录成功的情况下,运行这几个 JSP 页面,看看是什么效果?

8.4 小　　结

- 在 JSP/Servlet 中要实现过滤器,必须实现 Filter 接口,并在 web.xml 中定义部署过滤器,让服务器知道加载哪个过滤器。
- Filter 接口有 init()、doFilter()与 destroy() 3 个方法。这 3 个方法与 Servlet 接口

的 init()、service() 与 destroy() 类似。
- 过滤器必须在 web.xml 中部署,可以使用＜filter＞和＜filter-mapping＞标记部署过滤器。其中使用＜filter-name＞部署过滤器的名称,使用＜filter-class＞部署过滤器的类名,使用＜url-pattern＞部署 URL 的请求模式。

习 题 8

1. 简述过滤器的运行原理。
2. Filter 接口中有哪些方法?它们分别具有什么功能?
3. 在 web.xml 中部署过滤器需要哪些标记?这些标记的作用是什么?

第 9 章 EL 与 JSTL

本章主要内容

- 表达式语言(EL)
- JSP 标准标签库(JSTL)

在前面章节中编写 JSP 页面时,经常使用 Java 代码来实现页面显示逻辑。网页中夹杂着 HTML 与 Java 代码,给网页的设计与维护带来困难。我们可以使用 EL(Expression Language)来访问和处理应用程序的数据,也可以使用 JSTL(JavaServer Pages Standard Tag Library)来替换网页中实现页面显示逻辑的 Java 代码。这样 JSP 页面就尽量减少了 Java 代码的使用,为以后的维护提供了方便。

本章将重点介绍 EL 与 JSTL 核心标签库的基本用法。

9.1 使用 EL 访问对象的属性

9.1.1 核心知识

EL 是 JSP 2.0 规范中增加的,它的基本语法为:

${表达式}

类似于 JSP 表达式<%=表达式%>,EL 语句中的表达式值会被直接送到浏览器显示。使用 EL 可以获取对象的属性,如 JavaBean、数组或 List 对象。

1. 获取 JavaBean 的属性值

假设在 JSP 页面中有这样一句话。

```
<jsp:getProperty property="age" name="user"/>
```

那么,可以使用 EL 获取 user 的属性 age,修改如下:

```
${user.age}
```

其中,点运算符前面为 JavaBean 的对象 user,后面为该对象的属性 age,表示利用 user 对象的 getAge()方法取得值,而后显示在网页上。

2. 获取数组中的元素

假设在 JSP 页面中有这样一段话。

```
<%
    String dogs[]={"lili","huahua","guoguo"};
    request.setAttribute("array", dogs);
%>
```

那么,在页面某处可以使用 EL 取出数组中的元素,代码如下:

```
${array[0]}
${array[1]}
${array[2]}
```

3. 获取 List 对象中的元素

假设在 JSP 页面中有这样一段话。

```
<%
    ArrayList<UserBean>users=new ArrayList<UserBean>();
    UserBean ub1=new UserBean("zhang",20);
    UserBean ub2=new UserBean("zhao",50);
    users.add(ub1);
    users.add(ub2);
    request.setAttribute("array", users);
%>
```

其中,UserBean 有两个属性:name 和 age,那么在页面某处可以使用 EL 取出 UserBean 中的属性,代码如下:

```
${array[0].name}   ${array[0].age}
${array[1].name}   ${array[1].age}
```

9.1.2 能力目标

能够灵活使用 EL 表达式取出对象的属性。

9.1.3 任务驱动

1. 任务的主要内容

编写一个创建 JavaBean 对象的类 UserBean,该类中有两个属性(成员变量):name 和 age。然后,在 JSP 页面 example9_1.jsp 中使用 EL 取出该 JavaBean 对象的属性。

2. 任务的代码模板

将 example9_1.jsp 中的【代码】替换为 JSP 的代码。

UserBean.java

```
package bean;
public class UserBean {
    String name;
    int age;
```

```java
    public UserBean(){
        name="EL 表达式";
        age=5;
    }
    public String getName() {
        return name;
    }
    public void setName(String name) {
        this.name=name;
    }
    public int getAge() {
        return age;
    }
    public void setAge(int age) {
        this.age=age;
    }
}
```

example9_1.jsp

```jsp
<%@page language="java" contentType="text/html; charset=GBK" pageEncoding="GBK"%>
<%@page import="bean.*" %>
<html>
<head>
<title>EL 表达式</title>
</head>
<body>
    <jsp:useBean id="user" class="bean.UserBean" scope="page"/>
    姓名:【代码 1】    <!--使用 EL 取出 name 属性 -->
    <br>
    年龄:【代码 2】    <!--使用 EL 取出 age 属性 -->
</body>
</html>
```

3. 任务小结或知识扩展

1) EL 处理 null 值

对于 null 值直接以空字符串显示,而不是 null,运算时也不会发生错误或空指针异常。所以在使用 EL 访问对象的属性时,不需判断对象是否为 null 对象。这样就为编写程序提供了方便。

2) JSP 页面与 EL

JSP 页面默认支持 EL,但如果 JSP 页面使用 page 指令设置 isELIgnored 属性(默认为 false)值为 true 的话,则该页面不能使用 EL。

3) 对象的有效范围

在 EL 中,可以使用 EL 内置对象指定范围来访问属性,EL 内置对象将稍后介绍。如果不指定对象的有效范围,则以 page、request、session、application 的顺序查找 EL 中所指定的对象。

4）(.)与[]运算符的区别

EL中点运算符(.)和[]运算符,在一些情况下用法是一样的,总结如下:

(1)(.)运算符左边可以是JavaBean或Map对象。

(2)[]运算符左边可以是JavaBean、Map、数组或List对象。

使用EL如何取得Map对象中的值呢?假设在JSP页面中有这样一段话。

```
<%
    HashMap<String,String>map=new HashMap<String,String>();
    map.put("first","第一");
    map.put("second","第二");
    request.setAttribute("number", map);
%>
```

那么在页面某处可以使用EL获得Map中的值,代码如下:

```
${number.first}
${number.second}
```

或

```
${number["first"]}
${number["second"]}
```

4. 代码模板的参考答案

【代码1】:`${user.name}`
【代码2】:`${user.age}`

9.1.4 实践环节

如果把上述 example9_1.jsp 中的代码。

```
<jsp:useBean id="user" class="bean.UserBean" scope="page"/>
```

删掉,然后再运行程序,查看页面显示结果;如果不使用EL取值,而使用标记＜jsp:getProperty＞取值,那么在删掉

```
<jsp:useBean id="user" class="bean.UserBean" scope="page"/>
```

的情况下,运行网页会是什么结果?

9.2 EL内置对象

EL内置对象共有11个,本节中只是介绍几个常用的EL内置对象:pageScope、requestScope、sessionScope、applicationScope、param以及paramValues。

9.2.1 核心知识

1. 与作用范围相关的内置对象

与作用范围相关的EL内置对象有 pageScope、requestScope、sessionScope 和

applicationScope,分别可以获取 JSP 内置对象 pageContext、request、session 和 application 中的数据。如果在 EL 中没有使用内置对象指定作用范围,则从作用范围为 pageScope 的数据开始寻找。获取数据的格式如下:

${EL 内置对象.关键字对象.属性}

或

${EL 内置对象.关键字对象}

例如:

```
<jsp:useBean id="user" class="bean.UserBean" scope="page"/>
<jsp:setProperty name="user" property="name" value="EL 内置对象" />
name: ${pageScope.user.name}
```

再比如,在 JSP 页面中有这样一段话。

```
<%
    ArrayList<UserBean>users=new ArrayList<UserBean>();
    UserBean ub1=new UserBean("zhang",20);
    UserBean ub2=new UserBean("zhao",50);
    users.add(ub1);
    users.add(ub2);
    request.setAttribute("array", users);
%>
```

其中,UserBean 有两个属性:name 和 age,那么在 request 有效的范围内可以使用 EL 取出 UserBean 的属性,代码如下:

```
${requestScope.array[0].name}  ${requestScope.array[0].age}
${requestScope.array[1].name}  ${requestScope.array[1].age}
```

2. 与请求参数相关的内置对象

与请求参数相关的 EL 内置对象有 param 和 paramValues。获取数据的格式如下:

${EL 内置对象.参数名}

比如,input.jsp 的代码如下:

```
<form method="post" action="param.jsp">
    <p>姓名:<input type="text" name="username" size="15" /></p>
    <p>兴趣:
    <input type="checkbox" name="habit" value="看书"/>看书
    <input type="checkbox" name="habit" value="玩游戏"/>玩游戏
    <input type="checkbox" name="habit" value="旅游"/>旅游
    <p>
    <input type="submit" value="提交"/>
</form>
```

那么,在 param.jsp 页面中可以使用 EL 获取参数值,代码如下:

```
<%request.setCharacterEncoding("GBK");%>
```

```
<body>
<h2>EL 隐含对象 param、paramValues</h2>
姓名：${param.username}</br>
兴趣：
${paramValues.habit[0]}
${paramValues.habit[1]}
${paramValues.habit[2]}
```

9.2.2 能力目标

灵活使用 EL 内置对象从 JSP 内置对象中获取数据。

9.2.3 任务驱动

1. 任务的主要内容

编写一个 Servlet 类，在该类中使用 request 内置对象存储数据，然后从该 servlet 转发到 show.jsp 页面，最后在 show.jsp 页面中显示 request 内置对象的数据。首先运行 servlet，在 IE 地址栏中输入：

http://localhost:8080/ch9/saveServlet

程序运行结果如图 9.1 所示。

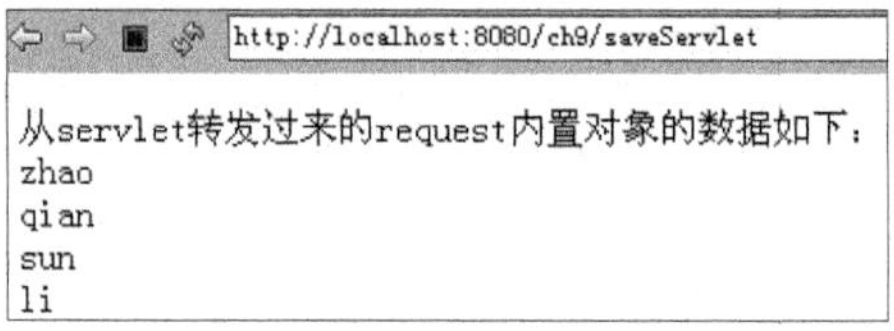

图 9.1 使用 EL 内置对象获取 JSP 内置对象的数据

2. 任务的代码模板

将 show.jsp 中的【代码】替换为 JSP 的代码。

web.xml

```
<servlet>
    <servlet-name>saveServlet</servlet-name>
    <servlet-class>servlet.SaveServlet</servlet-class>
</servlet>
<servlet-mapping>
    <servlet-name>saveServlet</servlet-name>
    <url-pattern>/saveServlet</url-pattern>
</servlet-mapping>
```

SaveServlet.java

```
package servlet;
import java.io.IOException;
import javax.servlet.RequestDispatcher;
import javax.servlet.ServletException;
import javax.servlet.http.HttpServlet;
import javax.servlet.http.HttpServletRequest;
import javax.servlet.http.HttpServletResponse;
public class SaveServlet extends HttpServlet {
    protected void doGet(HttpServletRequest request, HttpServletResponse response)
            throws ServletException, IOException {
        String names[]={"zhao","qian","sun","li"};
```

```
        request.setAttribute("name", names);
        RequestDispatcher dis=request.getRequestDispatcher("show.jsp");
        dis.forward(request, response);
    }
    protected void doPost(HttpServletRequest request,HttpServletResponse response)
            throws ServletException, IOException {
        doGet(request,response);
    }
}
```

show.jsp

```
<%@page language="java" contentType="text/html; charset=GBK" pageEncoding="GBK"%>
<html>
<head>
<title>EL内置对象</title>
</head>
<body>
从servlet转发过来的request内置对象的数据如下:<br>
    【代码1】<br><!--使用EL内置对象requestScope取出request中数组的第1个元素 -->
    【代码2】<br><!--使用EL内置对象requestScope取出request中数组的第2个元素 -->
    【代码3】<br><!--使用EL内置对象requestScope取出request中数组的第3个元素 -->
    【代码4】<br><!--使用EL内置对象requestScope取出request中数组的第4个元素 -->
</body>
</html>
```

3. 任务小结或知识扩展

EL内置对象与JSP内置对象不同,EL内置对象仅仅代表作用范围。

4. 代码模板的参考答案

【代码1】:${requestScope.name[0]}
【代码2】:${requestScope.name[1]}
【代码3】:${requestScope.name[2]}
【代码4】:${requestScope.name[3]}

9.2.4 实践环节

把任务中show.jsp页面里的代码

```
${requestScope.name[0]}<br>
${requestScope.name[1]}<br>
${requestScope.name[2]}<br>
${requestScope.name[3]}<br>
```

改成：

```
${name[0]}<br>
${name[1]}<br>
${name[2]}<br>
${name[3]}<br>
```

然后运行程序,查看运行结果。

9.3 基本输入输出标签

JSTL 是一个标准规范,但不在 JSP 的规范当中,所以需要下载 JSTL 实现(jar 包)。

可以登录网站 https://jstl.dev.java.net/下载 JSTL1.2 的 jar 包:jstl-impl-1.2.jar。另外,还需要 JSTL 标准接口与类(jstl.jar)。如果使用 Tomcat 作为 Web 服务器,可以在 Tomcat 的 webapps\examples\WEB-INF\lib 中找到 jstl.jar 文件。在 JSP 页面中要想使用 JSTL 核心标签库,必须把 jstl-impl-1.2.jar 与 jstl.jar 复制到 Web 工程的 WEB-INF\lib 中。同时在 JSP 页面中使用 taglib 标记定义前置名称与 uri 引用,代码如下:

```
<%@taglib prefix="c" uri="http://java.sun.com/jsp/jstl/core"%>
```

本书中只说明 JSTL 核心标签库中几个常用的标签,其他标签请参考 JSTL 说明文档或专门的书籍。

9.3.1 核心知识

1. <c:out> 标签

<c:out>用来显示数据的内容,与 <%= 表达式 %> 或 ${表达式}类似。格式如下:

```
<c:out value="输出的内容"[default="defaultValue"]/>
```

或

```
<c:out value="输出的内容">
    defaultValue
</c:out>
```

其中,value 值可以是一个 EL 表达式,也可以是一个字符串;default 可有可无,当 value 值不存在时,就输出 defaultValue。例如:

```
<c:out value="${param.data}" default="No Data" />
<br>
<c:out value="${param.nothing}" />
<br>
<c:out value="This is a String" />
```

输出的结果如图 9.2 所示。

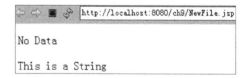

图 9.2 <c:out>标签

2. <c:set> 标签

(1) 设置作用域变量

可以使用<c:set>在 page、request、session、application 等范围内设置一个变量。格式如下:

```
< c: set  value =" value "  var =" varName "  [ scope =" page | request | session | application"]/>
```

将 value 值赋值给变量 varName。例如：

```
<c:set value="zhao" var="userName" scope="session"/>
```

相当于

```
<% session.setAttribute("userName","zhao"); %>
```

（2）设置 JavaBean 的属性

使用＜c:set＞设置 JavaBean 的属性时，必须使用 target 属性进行设置。格式如下：

```
<c:set value="value" target="target" property="propertyName"/>
```

将 value 赋值给 target 对象（JavaBean 对象）的 propertyName 属性。如果 target 为 null 或没有 set 方法则抛出异常。

3. ＜c:remove＞ 标签

如果要删除某个变量，则可以使用＜c:remove＞标签。例如：

```
<c:remove var="userName" scope="session"/>
```

相当于

```
<%session.removeAttribute("userName") %>
```

9.3.2 能力目标

掌握 JSTL 基本输入输出标签的使用方法。

9.3.3 任务驱动

1. 任务的主要内容

编写一个 JSP 页面 input_out.jsp，在该页面中使用＜c:set＞标签定义几个变量，并使用＜c:out＞标签输出这几个变量的值。运行效果如图 9.3 所示。

2. 任务的代码模板

将 input_out.jsp 中的【代码】替换为 JSP 的代码。

input_out.jsp

```
<%@page language="java" contentType=
"text/html; charset=GBK"
pageEncoding="GBK"%>
<%@taglib prefix="c" uri="http://java.sun.com/jsp/jstl/core"%>
<html>
<head>
<title>input_out.jsp</title>
</head>
```

图 9.3 ＜c:set＞与＜c:out＞标签

```
<body>
<c:set var="var1" value="setAndout" />
<c:set var="var2" value="1+2"/>
<c:set var="var3" value="${1+2}" />
<c:set var="var4"【代码】="request" value="${1 +2}" />
var1:<c:out value="${var1}" default="No Data" /><p>
var2:<c:out value="${var2}" default="No Data" /><p>
var3:<c:out value="${var3}" default="No Data" /><p>
var4:<c:out value="${var4+1}" default="No Data" /><p>
</body>
</html>
```

3. 任务小结或知识扩展

在9.2节中使用EL表达式就可以输出变量或表达式的值,那么我们为什么还要学习<c:out>标签呢？这不是多此一举吗？下面我们来猜猜这段程序的输出结果是什么？

```
<%
    String s="<p>有特殊字符</p>";
    request.setAttribute("exp", s);
%>
${exp}
```

运行时才发现其中的HTML标记<p>没有起到创建段落的作用。如果我们希望<p>达到创建段落的作用,那么必须把上面的语句

```
${exp}
```

改成：

```
<c:out value="<p>有特殊字符</p>" escapeXml="false" />
```

默认情况下,<c:out>将 <、>、'、"和 & 转换为 <、>、'、"和 &,如果不想转换,只需将 escapeXml 属性设置为 false。

4. 代码模板的参考答案

【代码】：scope

9.3.4 实践环节

把任务中 input_out.jsp 页面里的代码

```
var1:<c:out value="${var1}" default="No Data" /><p>
var2:<c:out value="${var2}" default="No Data" /><p>
var3:<c:out value="${var3}" default="No Data" /><p>
var4:<c:out value="${var4+1}" default="No Data" /><p>
```

改成：

```
var1:${var1}<p>
var2:${var2}<p>
var3:${var3}<p>
```

```
var4:${var4+1}<p>
```
然后再运行程序,查看结果有什么不同?

9.4 流程控制标签

9.4.1 核心知识

1. <c:if> 标签

<c:if>标签实现 if 语句的作用,具体语法格式如下:

```
<c:if test="条件表达式">
    主体内容
</c:if>
```

其中,条件表达式可以是 EL 表达式,也可以是 JSP 表达式。如果表达式的值为 true,则执行<c:if>的主体内容,但是没有相对应的<c:else>标签。如果想在条件成立时执行一块内容,不成立时执行另一块内容,则可以使用<c:choose>、<c:when>及<c:otherwise>标签。

2. <c:choose>、<c:when>及<c:otherwise> 标签

<c:choose>、<c:when>及<c:otherwise>标签实现 if/elseif/else 语句的作用。具体语法格式如下:

```
<c:choose>
    <c:when test="条件表达式 1">
        主体内容 1
    </c:when>
    <c:when test="条件表达式 2">
        主体内容 2
    </c:when>
    <c:otherwise>
        表达式都不正确时,执行的主体内容
    </c:otherwise>
</c:choose>
```

9.4.2 能力目标

掌握<c:if>、<c:choose>、<c:when>及<c:otherwise>标签的使用方法。

9.4.3 任务驱动

1. 任务的主要内容

编写一个 JSP 页面 ifelse.jsp,在该页面中使用<c:set>标签把两个字符串设置在 request 内置对象中。使用<c:if>标签求出这两个字符串的最大值(按字典顺序比较大小),使用<c:choose>、<c:when>及<c:otherwise>标签求出这两个字符串的最小值。

2. 任务的代码模板

将 ifelse.jsp 中的【代码】替换为 JSP 的代码。

ifelse.jsp

```jsp
<%@page language="java" contentType="text/html; charset=GBK"
    pageEncoding="GBK"%>
<%@taglib prefix="c" uri="http://java.sun.com/jsp/jstl/core"%>
<html>
<head>
<meta http-equiv="Content-Type" content="text/html; charset=ISO-8859-1">
<title>ifelse.jsp</title>
</head>
<body>
<c:set value="if" var="firstNumber" scope="request"/>
<c:set value="else" var="secondNumber" scope="request"/>
<c:if 【代码 1】>
    最大值为${firstNumber}
</c:if>
<c:if 【代码 2】>
    最大值为${secondNumber}
</c:if>
<c:choose>
    <c:when 【代码 3】>
        最小值为${firstNumber}
    </c:when>
    <c:otherwise>
        最小值为${secondNumber}
    </c:otherwise>
</c:choose>
</body>
</html>
```

3. 任务小结或知识扩展

　　＜c:when＞及＜c:otherwise＞必须放在＜c:choose＞中。当＜c:when＞的 test 结果为 true 时，会输出＜c:when＞的主体内容，而不理会＜c:otherwise＞的内容。＜c:choose＞中可有多个＜c:when＞，程序会从上到下进行条件判断，如果有个＜c:when＞的 test 结果为 true，就输出其主体内容，之后的＜c:when＞就不再执行。如果所有的＜c:when＞的 test 结果都为 false，则会输出＜c:otherwise＞的内容。＜c:if＞与＜c:choose＞也可以嵌套使用，例如：

```jsp
<c:set value="fda" var="firstNumber" scope="request"/>
<c:set value="else" var="secondNumber" scope="request"/>
<c:set value="ddd" var="threeNumber" scope="request"/>

<c:if test="${firstNumber>secondNumber}">
    <c:choose>
        <c:when test="${firstNumber<threeNumber}">
            最大值为${threeNumber}
        </c:when>
        <c:otherwise>
            最大值为${firstNumber}
```

```
            </c:otherwise>
        </c:choose>
</c:if>

<c:if test="${secondNumber>firstNumber}">
    <c:choose>
        <c:when test="${secondNumber<threeNumber}">
            最大值为${threeNumber}
        </c:when>
        <c:otherwise>
            最大值为${secondNumber}
        </c:otherwise>
    </c:choose>
</c:if>
```

4. 代码模板的参考答案

【代码 1】:test="${firstNumber>secondNumber}"
【代码 2】:test="${firstNumber<secondNumber}"
【代码 3】:test="${firstNumber<secondNumber}"

9.4.4　实践环节

编写一个 JSP 页面 pratice9_4.jsp,在该页面中使用＜c:set＞标签把 3 个字符串设置在 request 内置对象中。使用＜c:if＞、＜c:when＞、＜c:otherwise＞和＜c:choose＞标签求出这 3 个字符串的最小值。

9.5　迭 代 标 签

9.5.1　核心知识

JSTL 的＜c:forEach＞标签可以实现程序中的 for 循环。语法格式如下:

```
<c:forEach var="变量名" items="数组或 Collection 对象">
    循环体
</c:forEach>
```

其中,items 属性可以是数组或 Collection 对象,每次循环读取对象中的一个元素,并赋值给 var 属性指定的变量,之后就可以在循环体使用 var 指定的变量获取对象的元素。例如,在 JSP 页面中有这样一段代码。

```
<%
    ArrayList<UserBean>users=new ArrayList<UserBean>();
    UserBean ub1=new UserBean("zhao",20);
    UserBean ub2=new UserBean("qian",40);
    UserBean ub3=new UserBean("sun",60);
    UserBean ub4=new UserBean("li",80);
    users.add(ub1);
    users.add(ub2);
```

```
            users.add(ub3);
            users.add(ub4);
            request.setAttribute("usersKey", users);
    %>
```

那么,在页面某处我们可以使用<c:forEach>循环遍历出数组中的元素。代码如下:

```
<table>
    <tr>
        <th>姓名</th>
        <th>年龄</th>
    </tr>
<c:forEach var="user" items="${requestScope.usersKey}">
    <tr>
        <td>${user.name}</td>
        <td>${user.age}</td>
    </tr>
</c:forEach>
</table>
```

9.5.2 能力目标

能够灵活使用<c:forEach>标签循环遍历集合中的元素。

9.5.3 任务驱动

1. 任务的主要内容

把 7.2 节任务驱动中 showAllGoods.jsp 页面里的 for 语句改成<c:forEach>标签。

2. 任务的代码模板

将 showAllGoods.jsp 中的【代码】替换为 JSP 的代码。

showAllGoods.jsp

```
<%@page language="java" contentType="text/html; charset=GBK" pageEncoding="GBK"%>
<%@taglib prefix="c" uri="http://java.sun.com/jsp/jstl/core"%>
<%@page import="java.util.ArrayList" %>
<%@page import="bean.Goods" %>
<html>
<head>
<title>showAllGoods.jsp</title>
</head>
<body>
    <table border="1">
        <tr>
            <th>商品编号</th>
            <th>商品名称</th>
            <th>商品价格</th>
            <th>商品类别</th>
        </tr>
```

```
<c:forEach【代码 1】【代码 2】>
    <tr>
        <td>${good.goodsId}</td>
        <td>${good.goodsName}</td>
        <td>${good.goodsPrice}</td>
        <td>${good.goodsType}</td>
    </tr>
</c:forEach>
</table>
</body>
</html>
```

3. 任务小结或知识扩展

有些情况下,我们需要为<c:forEach>标签里的 var 属性指定初始值(begin)、结束值(end)和步长(step)。例如:

```
<table border=1>
    <tr>
        <th>Value</th><th>Square</th>
    </tr>
<c:forEach var="x" begin="0" end="10" step="2">
    <tr>
        <td><c:out value="${x}"/></td>
        <td><c:out value="${x * x}"/></td>
    </tr>
</c:forEach>
</table>
```

上述程序运行结果如图 9.4 所示。

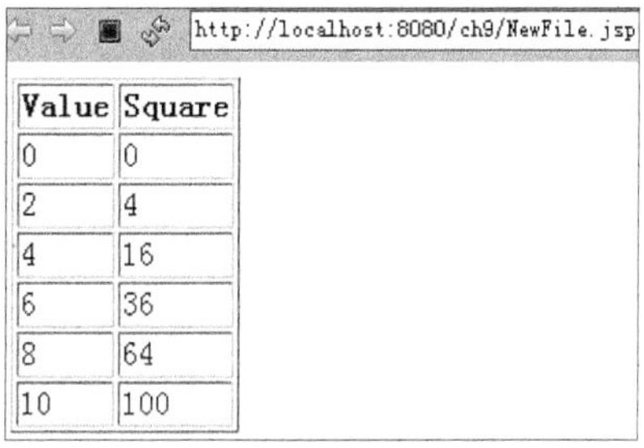

4. 代码模板的参考答案

【代码 1】:var="good"
【代码 2】:items="${requestScope.goods}"

图 9.4　<c:forEach>标签

9.5.4　实践环节

编写一个 JSP 页面 pratice9_5.jsp,在该页面中使用<c:forEach>标签输出九九乘法表。

9.6　小　　结

- 在 JSP 页面中一些简单的属性、请求参数等值的获取,一些简单的运算或判断,可以使用 EL 表达式来处理,减少了页面中的 Java 代码。
- 可以使用 EL 表达式获取对象的属性值,比如,JavaBean 对象、Map 对象、数组或 List 对象,还可以使用它获取 JSP 内置对象中的数据。
- 在 JSP 页面中可以使用 JSTL 标签来替代实现页面逻辑的 Java 程序,比如,<c:if>替代 if 语句,<c:forEach>替代 for 语句。

习 题 9

1. 在 Web 应用程序中有以下程序代码段,执行后转发到某个 JSP 页面。

```
ArrayList<String>  dogNames=new ArrayList<String>();
dogNames.add("goodDog");
request.setAttribute("dogs", dogNames);
```

以下( )选项可以正确地使用 EL 取得数组中的值。

    A. ${dogs.0}　　B. ${dogs[0]}　　C. ${dogs.[0]}　　D. ${dogs"0"}

2. ( )JSTL 标签可以实现 Java 程序中的 if 语句功能。

    A. <c:set>　　B. <c:out>　　C. <c:forEach>　　D. <c:if>

3. ( )不是 EL 的内置对象。

    A. request　　　　　　　　　　B. pageScope

    C. sessionScope　　　　　　　　D. applicationScope

4. ( )JSTL 标签可以实现 Java 程序中的 for 语句功能。

    A. <c:set>　　B. <c:out>　　C. <c:forEach>　　D. <c:if>

第 10 章 地址簿管理信息系统

本章通过一个小型的地址簿管理信息系统,讲述如何采用 JSP+JavaBean+Servlet 的模式来开发一个 Web 应用。系统将业务逻辑封装在 JavaBean 中,使系统的可维护性和可扩展性大为提高。

系统的开发环境如下。
- 操作系统:Windows XP SP2。
- 数据库:Oracle 10g。
- JSP 引擎:Tomcat 6.0。
- 集成开发环境(IDE):Eclipse IDE for Java EE Developers。

10.1 系统设计

10.1.1 系统功能需求

地址簿管理信息系统是针对注册用户使用的系统。系统提供的功能如下:
(1) 非注册用户可以注册为注册用户。
(2) 成功注册的用户,可以登录系统。
(3) 成功登录的用户,可以添加、修改、删除以及浏览自己的朋友信息。
(4) 成功登录的用户,可以修改自己的登录密码。

10.1.2 系统模块划分

注册用户使用地址簿管理信息系统可以添加、修改、删除以及查询自己的朋友信息,具体的系统功能模块如下。

1. 用户注册

新用户填写注册信息,包括用户名、密码和确认密码。输入用户名时,系统会提示用户名是否可以使用。

2. 用户登录

用户输入用户名、密码进行登录。登录失败,系统回到登录画面继续登录。登录成功,进入系统管理主页面(main.jsp),包括添加朋友信息、修改朋友信息、删除朋友信息、查询朋友信息、修改密码以及退出系统等功能。

3. 添加朋友信息

用户填写朋友信息表单,包括朋友姓名、生日、电话、E-mail、地址以及关系等信息。提交朋友信息表单时,系统使用 JavaScript 验证信息是否输入以及信息格式是否合法。

4. 修改朋友信息

系统首先根据成功登录的用户名查询出该用户的所有朋友信息,然后用户选择某个朋友进行信息修改。

5. 删除朋友信息

系统首先根据成功登录的用户名查询出该用户的所有朋友信息,然后用户选择某个或多个朋友进行删除。

6. 查询朋友信息

系统根据成功登录的用户名查询出该用户的所有朋友信息。

7. 修改密码

成功登录的用户,从主页面进入该页面修改自己的密码。

8. 退出系统

成功登录的用户,在主页面点击退出系统链接,系统首先清除用户的会话(session),然后回到登录页面。

10.2 数据库设计

系统采用加载纯 Java 数据库驱动程序的方式连接 Oracle 10g 数据库。在 Oracle 10g 的默认数据库 orcl 中创建两张表:usertable 与 friendinfo。

10.2.1 数据库概念结构设计

根据系统设计与分析,可以设计出如下数据结构。

1. 用户信息

包括用户名和密码等,一个用户可以添加多个朋友信息。

2. 朋友信息

包括朋友 ID、姓名、生日、电话、E-mail、地址、关系以及所属的用户等信息。

根据以上的数据结构,结合数据库设计的特点,可以画出如图 10.1 所示的数据库概念结构图。

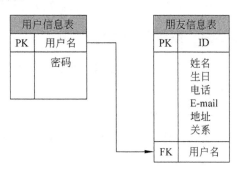

图 10.1 数据库概念结构图

10.2.2 数据库逻辑结构设计

将数据库概念结构图转换为 Oracle 数据库所支持的实际数据模型,即数据库的逻辑结构。系统中两张表的设计如表 10.1 和表 10.2 所示。

表 10.1 用户信息表（usertable）

字 段	含 义	类 型	长 度	是否为空
userName	用户名	varchar	20	no
password	密码	varchar	20	no

表 10.2 朋友信息表（friendinfo）

字 段	含 义	类 型	长 度	是否为空
id	朋友编号	varchar	17	no
name	朋友姓名	varchar	20	no
birthday	朋友生日	date	—	yes
telephone	朋友电话	varchar	20	yes
email	朋友 E-mail	varchar	20	yes
address	朋友地址	varchar	50	yes
relation	关系	varchar	10	no
userName	用户名	varchar	20	no

10.2.3 创建数据表

根据数据库的逻辑结构，创建数据表的代码如下：

```
drop table friendinfo;
drop table usertable;
create table usertable (
    userName varchar(20) not null,
    password varchar(20) not null,
    constraint pk_usertable primary key (userName)
);
create table friendinfo (
    id varchar(17) not null,
    name varchar(20) not null,
    birthday date null,
    telephone varchar(20) null,
    email varchar(20) null,
    address varchar(50)  null,
    relation varchar(10) not null,
    userName varchar(20) not null,
    constraint pk_friendinfo primary key (id),
        constraint fk_friendinfo_1 foreign key (userName)
            references usertable (userName)
);
commit;
```

10.3 系统管理

10.3.1 导入相关的 jar 包

在本章中新建一个 Web 工程 ch10,在 ch10 工程中开发本系统。由于在本系统中所有 JSP 页面尽量使用 EL 表达式和 JSTL 标签,又因为本系统采用纯 Java 数据库驱动程序连接 Oracle 10g 数据库。所以,我们需要把 jstl-impl-1.2.jar、jstl.jar 和 classes12.jar 复制到 ch10/WEB-INF/lib 文件夹中。具体操作步骤请参考本书中的 5.2 节和 9.3 节。

10.3.2 JSP 页面管理

本系统中用到的所有 JSP 页面(包括相关的 JavaScript 文件)都保存在 ch10 的 WebContent 目录下。系统中有些 JSP 页面使用到 CSS 与 JavaScript。有关 CSS 与 JavaScript 的知识,超出了本书的讨论范围,请读者查阅相关的书籍。

本系统中除了登录、注册以及修改密码等操作不在系统管理主页面(main.jsp)实现,别的操作都在 main.jsp 中实现。在该页面中使用 DIV+CSS+Iframe 进行布局管理。

用户首选应在浏览器地址栏中输入 http://localhost:8080/ch10/login.jsp 访问登录页面,登录成功,进入系统管理主页面,main.jsp 运行效果如图 10.2 所示。

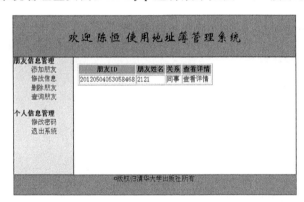

图 10.2 系统管理主页面

main.jsp

```
<%@page language="java" contentType="text/html; charset=GBK" pageEncoding="GBK"%>
<%@page isELIgnored="false" %>
<%@taglib prefix="c" uri="http://java.sun.com/jsp/jstl/core"%>
<%
    String path=request.getContextPath();
    String basePath=request.getScheme() +"://"
        +request.getServerName() +":" +request.getServerPort()
        +path +"/";
%>
<html>
<head>
```

```
<base href="<%=basePath%>">
<title>管理信息主页面</title>
<!--CSS 代码 -->
<style type="text/css">
# top {
    position: absolute;
    top: 0px;
    left: 200px;
    right: auto;
    background-image: url(images/Bluehills.jpg);
    bottom: auto;
    background-color: SkyBlue;
    border: 1px solid # ff3399;
    width: 700px;
    height: 100px;
}

# left {
    float: left;
    position: absolute;
    top: 100px;
    left: 200px;
    right: auto;
    bottom: auto;
    background-color: LightCyan;
    border: 1px solid # 9400D3;
    width: 150px;
    height: 400px;
}

# right1 {
    float: right;
    position: absolute;
    top: 100px;
    left: 350px;
    right: auto;
    bottom: auto;
    background-color: # FFB6C1;
    border: 1px solid # EE1289;
    width: 550px;
    height: 400px;
}

# bottom {
    position: absolute;
    top: 500px;
    left: 200px;
    right: auto;
    bottom: auto;
    background-color: SkyBlue;
    border: 1px solid # DAA520;
```

```
        width: 700px;
        height: 50px;
}
</style>
</head>
<!--Javascript 代码 -->
<script type="text/javaScript">
    function showHide(objname) {
        //只对主菜单设置 cookie
        var obj=document.getElementById(objname);
        if (objname.indexOf('_1') <0 || objname.indexOf('_10') >0) {
            if (obj.style.display =='block' || obj.style.display =='')
                obj.style.display='none';
            else
                obj.style.display='block';
            return true;
        }
        //正常设置 cookie
        var ckstr=getCookie('menuitems');
        var ckstrs=null;
        var okstr='';
        var ischange=false;
        if (ckstr ==null)
            ckstr='';
        ckstrs=ckstr.split(',');
        objname=objname.replace('items', '');
        if (obj.style.display =='block' || obj.style.display =='') {
            obj.style.display='none';
            for ( var i=0; i <ckstrs.length; i++) {
                if (ckstrs[i] =='')
                    continue;
                if (ckstrs[i] ==objname) {
                    ischange=true;
                } else
                    okstr +=(okstr =='' ? ckstrs[i] : ',' +ckstrs[i]);
            }
            if (ischange)
                setCookie('menuitems', okstr, 7);
        } else {
            obj.style.display='block';
            ischange=true;
            for ( var i=0; i <ckstrs.length; i++) {
                if (ckstrs[i] ==objname) {
                    ischange=false;
                    break;
                }
            }
            if (ischange) {
                ckstr= (ckstr ==null ? objname : ckstr +',' +objname);
                setCookie('menuitems', ckstr, 7);
            }
```

```
            }
        }
        //读写cookie函数
        function getCookie(c_name) {
            if (document.cookie.length >0) {
                c_start=document.cookie.indexOf(c_name +"=")
                if (c_start !=-1) {
                    c_start=c_start +c_name.length +1;
                    c_end=document.cookie.indexOf(";", c_start);
                    if (c_end ==-1) {
                        c_end=document.cookie.length;
                    }
                    return unescape(document.cookie.substring(c_start, c_end));
                }
            }
            return null
        }
        function setCookie(c_name, value, expiredays) {
            var exdate=new Date();
            exdate.setDate(exdate.getDate() +expiredays);
            document.cookie=c_name
                    +"="
                    +escape(value)
                    + ((expiredays ==null) ? "" : ";expires="
                        +exdate.toGMTString());          //使设置的有效时间正确
        }
</script>
<body>
    <div id="top">
        <br>
        <br>
        <center>
        <font size=6 face=华文新魏>欢迎 ${sessionScope.user.uname} 使用地址簿管
        理系统</font>
        </center>
    </div>

    <div id="left">
        <dl>
            <dt onClick='showHide("items1_1")'>
                <b>朋友信息管理</b>
            </dt>
            <dd style='display: block' id='items1_1'>
                <table>
                    <tr>
                        <td><a href="addFriend.jsp" target="right"
                            style="text-decoration: none">添加朋友</a>
                        </td>
                    </tr>
                    <tr>
                        <td><a href="queryFriend? flag=update" target="right"
```

```html
                        style="text-decoration: none">修改信息</a>
                </td>
            </tr>

            <tr>
                <td><a href="queryFriend? flag=del" target="right"
                        style="text-decoration: none">删除朋友</a>
                </td>
            </tr>

            <tr>
                <td><a href="queryFriend" target="right"
                        style="text-decoration: none">查询朋友</a>
                </td>
            </tr>

        </table>
    </dd>
</dl>
<dl>
    <dt onClick='showHide("items1_2")'>
        <b>个人信息管理</b>
    </dt>
    <dd style='display: block' id='items1_2'>
        <table>
            <tr>
                <td><a href="upadatepassword.jsp"
                        style="text-decoration: none">修改密码</a>
                </td>
            </tr>

            <tr>
                <td><a href="exitUser"
                        style="text-decoration: none">退出系统</a>
                </td>
            </tr>
        </table>
    </dd>
</dl>
</div>

<div id="right1">
    <iframe src="queryFriend" name="right" width="100%" height="100%"
        frameborder="0"></iframe>
</div>

<div id="bottom">
    <center>&copy;版权归清华大学出版社所有</center>
</div>
</body>
</html>
```

10.3.3 组件与 servlet 管理

本系统中使用的组件与 servlet 包层次结构如图 10.3 所示。

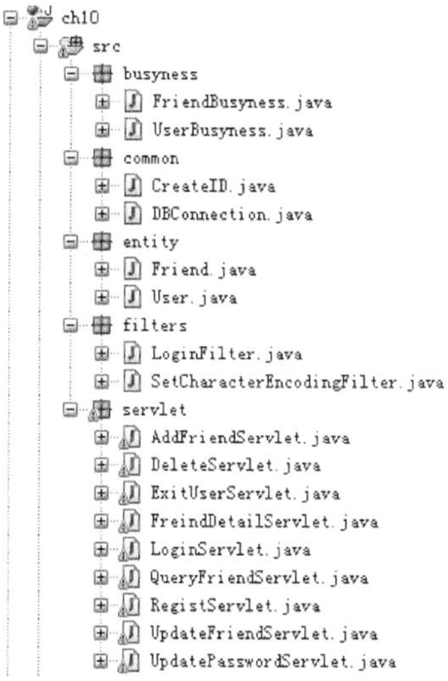

1. busyness 包

图 10.3 中 busyness 包里存放的 Java 程序都是实现业务逻辑处理的 bean(业务模型)，包括用户注册、登录、朋友信息(增、删、改、查)等业务的处理。

2. common 包

该包中的 CreateID 类是实现朋友 ID 主键的产生，DBConnection 类是实现数据的连接与关闭等操作。

3. entity 包

该包中的 Friend 类和 User 类是实现数据封装的实体 bean(实体模型)。

图 10.3 包层次结构图

4. filters 包

系统中使用过滤器(filter)解决中文乱码和登录验证等问题，系统中过滤器 Java 代码的实现都存放在 filters 包中。

5. servlet 包

系统中控制器的实现都存放在 servlet 包中。

10.3.4 配置文件管理

在 ch10 的配置文件 web.xml 中，部署了本系统用到的过滤器与控制器。web.xml 的代码实现如下：

web.xml

```
<?xml version="1.0" encoding="UTF-8"?>
<web-app>
    <servlet>
        <servlet-name>registServlet</servlet-name>
        <servlet-class>servlet.RegistServlet</servlet-class>
    </servlet>
    <servlet-mapping>
        <servlet-name>registServlet</servlet-name>
        <url-pattern>/registServlet</url-pattern>
    </servlet-mapping>

    <servlet>
        <servlet-name>loginServlet</servlet-name>
        <servlet-class>servlet.LoginServlet</servlet-class>
    </servlet>
    <servlet-mapping>
        <servlet-name>loginServlet</servlet-name>
```

```xml
        <url-pattern>/login</url-pattern>
    </servlet-mapping>

    <servlet>
        <servlet-name>addFriendServlet</servlet-name>
        <servlet-class>servlet.AddFriendServlet</servlet-class>
    </servlet>
    <servlet-mapping>
        <servlet-name>addFriendServlet</servlet-name>
        <url-pattern>/addFriend</url-pattern>
    </servlet-mapping>

    <servlet>
        <servlet-name>queryFriendServlet</servlet-name>
        <servlet-class>servlet.QueryFriendServlet</servlet-class>
    </servlet>
    <servlet-mapping>
        <servlet-name>queryFriendServlet</servlet-name>
        <url-pattern>/queryFriend</url-pattern>
    </servlet-mapping>

    <servlet>
        <servlet-name>freindDetailServlet</servlet-name>
        <servlet-class>servlet.FreindDetailServlet</servlet-class>
    </servlet>
    <servlet-mapping>
        <servlet-name>freindDetailServlet</servlet-name>
        <url-pattern>/freindDetail</url-pattern>
    </servlet-mapping>

    <servlet>
        <servlet-name>deleteServlet</servlet-name>
        <servlet-class>servlet.DeleteServlet</servlet-class>
    </servlet>
    <servlet-mapping>
        <servlet-name>deleteServlet</servlet-name>
        <url-pattern>/delete</url-pattern>
    </servlet-mapping>

    <servlet>
        <servlet-name>updateFriendServlet</servlet-name>
        <servlet-class>servlet.UpdateFriendServlet</servlet-class>
    </servlet>
    <servlet-mapping>
        <servlet-name>updateFriendServlet</servlet-name>
        <url-pattern>/updateFriend</url-pattern>
    </servlet-mapping>

    <servlet>
        <servlet-name>updatePasswordServlet</servlet-name>
        <servlet-class>servlet.UpdatePasswordServlet</servlet-class>
```

```xml
</servlet>
<servlet-mapping>
    <servlet-name>updatePasswordServlet</servlet-name>
    <url-pattern>/updatePassword</url-pattern>
</servlet-mapping>

<servlet>
    <servlet-name>exitUserServlet</servlet-name>
    <servlet-class>servlet.ExitUserServlet</servlet-class>
</servlet>
<servlet-mapping>
    <servlet-name>exitUserServlet</servlet-name>
    <url-pattern>/exitUser</url-pattern>
</servlet-mapping>

<filter>
    <filter-name>setCharacterEncoding</filter-name>
    <filter-class>filters.SetCharacterEncodingFilter</filter-class>
    <init-param>
        <param-name>encoding</param-name>
        <param-value>GBK</param-value>
    </init-param>
</filter>
<filter-mapping>
    <filter-name>setCharacterEncoding</filter-name>
    <url-pattern>/*</url-pattern>
</filter-mapping>

<filter>
    <filter-name>loginFilter</filter-name>
    <filter-class>filters.LoginFilter</filter-class>
    <init-param>
        <param-name>LOGON_URI</param-name>
        <param-value>/login.jsp</param-value>
    </init-param>
    <init-param>
        <param-name>LOGON_SERVLET</param-name>
        <param-value>/login</param-value>
    </init-param>
    <init-param>
        <param-name>Regist_SERVLET</param-name>
        <param-value>/registServlet</param-value>
    </init-param>
    <init-param>
        <param-name>Regist_URI</param-name>
        <param-value>/regist.jsp</param-value>
    </init-param>
</filter>
<filter-mapping>
    <filter-name>loginFilter</filter-name>
    <url-pattern>/*</url-pattern>
```

```
        </filter-mapping>
</web-app>
```

10.4 组件设计

本系统中使用到的组件有过滤器、数据库连接与关闭、实体模型（数据封装 bean）和业务模型（业务处理 bean）4 部分。

10.4.1 过滤器

本系统中使用了两个过滤器：设置字符编码过滤器与登录验证过滤器。有关过滤器的知识，请读者参考第 8 章的内容。

1. 设置字符编码过滤器

在系统中使用过滤器解决中文乱码的问题。当用户提交请求时，在请求处理之前，系统使用过滤器把用户提交的信息进行解码与编码，避免了乱码的出现。具体代码如下：

SetCharacterEncodingFilter.java

```
package filters;
import java.io.IOException;
import javax.servlet.Filter;
import javax.servlet.FilterChain;
import javax.servlet.FilterConfig;
import javax.servlet.ServletException;
import javax.servlet.ServletRequest;
import javax.servlet.ServletResponse;
//过滤器设置字符编码为 GBK
public class SetCharacterEncodingFilter implements Filter {
    protected String encoding=null;
    public void destroy() {
        this.encoding=null;
    }
    public void doFilter(ServletRequest request, ServletResponse response,
        FilterChain chain)
    throws IOException, ServletException {
        String encoding=this.encoding;
        if (encoding !=null)
            request.setCharacterEncoding(encoding);
        //执行下一个过滤器
        chain.doFilter(request, response);
    }
    public void init(FilterConfig filterConfig) throws ServletException {
        this.encoding=filterConfig.getInitParameter("encoding");
    }
}
```

2. 登录验证过滤器

从系统分析得知，只有用户成功登录了，才能使用该系统。也就是说，用户不能成功登录时,不允许使用除登录、注册之外的功能。当用户通过 URL 请求时,系统首先使用过滤器判别用户访问的是不是登录或注册的功能,如果不是,就判断用户的会话(session)是否存在,如果不存在,就提示用户先登录。具体代码如下:

LoginFilter.java

```java
package filters;
import java.io.IOException;
import java.io.PrintWriter;
import javax.servlet.Filter;
import javax.servlet.FilterChain;
import javax.servlet.FilterConfig;
import javax.servlet.ServletException;
import javax.servlet.ServletRequest;
import javax.servlet.ServletResponse;
import javax.servlet.http.HttpServletRequest;
import javax.servlet.http.HttpServletResponse;
import javax.servlet.http.HttpSession;
public class LoginFilter implements Filter {
    private static final String LOGON_URI="LOGON_URI";
    private static final String Regist_URI="Regist_URI";
    private static final String LOGON_SERVLET="LOGON_SERVLET";
    private static final String Regist_SERVLET="Regist_SERVLET";
    private String logon_page;
    private String regist_page;
    private String logon_servlet;
    private String regist_servlet;
    public void destroy() {
    }
    public void doFilter(ServletRequest request, ServletResponse response,
            FilterChain chain) throws IOException, ServletException {
        HttpServletRequest req=(HttpServletRequest) request;
        HttpServletResponse resp=(HttpServletResponse) response;
        resp.setContentType("text/html;");
        resp.setCharacterEncoding("GBK");
        HttpSession session=req.getSession();
        PrintWriter out=resp.getWriter();
        //得到用户请求的 URI
        String request_uri=req.getRequestURI();
        //得到 Web 应用程序的上下文路径
        String ctxPath=req.getContextPath();
        //去除上下文路径,得到剩余部分的路径
        String uri=request_uri.substring(ctxPath.length());
        //判断用户访问的是否是登录页面或注册页面
        if (uri.equals(logon_page)||uri.equals(regist_page)
                ||uri.equals(logon_servlet)||uri.equals(regist_servlet)) {
            //执行下一个过滤器
            chain.doFilter(request, response);
        } else {
            //如果访问的不是登录页面,则判断用户是否已经登录
            if (null !=session.getAttribute("user")
```

```
                && ""!=session.getAttribute("user")) {
                chain.doFilter(request, response);
            } else {
                String path=ctxPath+logon_page;
                out.println("您没有登录,请先<a href="+pah+" target=_top>登录</a>!");
                return;
            }
        }
    }
    public void init(FilterConfig config) throws ServletException {
        //从 web.xml 的部署描述符中获取初始参数值
        logon_page=config.getInitParameter(LOGON_URI);
        regist_page=config.getInitParameter(Regist_URI);
        logon_servlet=config.getInitParameter(LOGON_SERVLET);
        regist_servlet=config.getInitParameter(Regist_SERVLET);
    }
}
```

10.4.2 数据库连接与关闭

1. 主键生成器

本系统中朋友信息表的主键(ID)产生的策略是:格式化系统时间,主键格式(共 17 位)如下:

yyyyMMddHHmmssSSS

主键生成器的代码如下:

CreateID.java

```
package common;
import java.util.Date;
import java.text.SimpleDateFormat;
public class CreateID {
    //主键产生器
    public static String getStringID(){
        String id=null;
        Date date=new Date();
        SimpleDateFormat sdf=new SimpleDateFormat("yyyyMMddHHmmssSSS");
        id=sdf.format(date);
        return id;
    }
}
```

2. 数据库的连接与关闭

数据库的连接与关闭是由 common 包中的 DBConnection 类实现的。类中的方法功能说明如下。

(1) public synchronized static Connection getOneCon():从连接池中获得一个连接对象。

（2）public static void close(ResultSet rs)：关闭结果集对象。

（3）public static void close(PreparedStatement ps)：关闭预处理对象。

（4）public synchronized static void close(Connection con)：把连接对象放回连接池中。

DBConnection 类的代码如下：

DBConnection.java

```java
package common;
import java.sql.*;
import java.util.ArrayList;
public class DBConnection {
    //存放 Connection 对象的数组,数组被看成连接池
    static ArrayList<Connection>list=new ArrayList<Connection>();
    //从连接池中取出一个连接对象
    public synchronized static  Connection getOneCon(){
        //如果连接池中有连接对象
        if(list.size()>0){
            return list.remove(0);
        }
        //连接池没有连接对象创建连接放到连接池中
        else{
            for(int i=0;i<5;i++){
                try {
                    Class.forName("oracle.jdbc.driver.OracleDriver");
                    Connection con=DriverManager.
                        getConnection("jdbc:oracle:thin:@localhost:1521:
                        orcl","system","system");
                    list.add(con);
                } catch (Exception e) {
                    //TODO Auto-generated catch block
                    e.printStackTrace();
                }
            }
            return list.remove(0);
        }
    }
    //关闭结果集对象
    public static void close(ResultSet rs){
        try {
            if(rs!=null)
                rs.close();
        } catch (SQLException e) {
            //TODO Auto-generated catch block
            e.printStackTrace();
        }
    }
    //关闭预处理语句
    public static void close(PreparedStatement ps){
        try {
            if(ps!=null)
                ps.close();
```

```
            } catch (SQLException e) {
                //TODO Auto-generated catch block
                e.printStackTrace();
            }
        }
        //把连接对象放回连接池中
        public synchronized static void close(Connection con){
            if(con!=null)
                list.add(con);
        }
    }
```

10.4.3 实体模型

实体模型主要用于封装 JSP 页面提交的信息以及与数据库交互的数据传递。本系统中共使用两个实体模型：用户（User）和朋友（Friend）。具体实现如下：

Friend.java

```
package entity;
public class Friend {
    private String id;
    private String name;
    private String birthday;
    private String telephone;
    private String email;
    private String address;
    private String relation;
    public Friend(){
        
    }
    public String getId() {
        return id;
    }
    public void setId(String id) {
        this.id=id;
    }
    public String getName() {
        return name;
    }
    public void setName(String name) {
        this.name=name;
    }
    public String getBirthday() {
        return birthday;
    }
    public void setBirthday(String birthday) {
        this.birthday=birthday;
    }
    public String getTelephone() {
        return telephone;
```

```java
    }
    public void setTelephone(String telephone) {
        this.telephone=telephone;
    }
    public String getEmail() {
        return email;
    }
    public void setEmail(String email) {
        this.email=email;
    }
    public String getAddress() {
        return address;
    }
    public void setAddress(String address) {
        this.address=address;
    }
    public String getRelation() {
        return relation;
    }
    public void setRelation(String relation) {
        this.relation=relation;
    }
}
```

User.java

```java
package entity;
public class User {
    private String uname;
    private String upass;
    public User(){
    }
    public String getUname() {
        return uname;
    }
    public void setUname(String uname) {
        this.uname=uname;
    }
    public String getUpass() {
        return upass;
    }
    public void setUpass(String upass) {
        this.upass=upass;
    }
}
```

10.4.4 业务模型

系统中共用到两个业务 bean：FriendBusyness 和 UserBusyness。和用户有关的业务逻辑处理都写在 UserBusyness 类里，和朋友信息有关的业务逻辑处理都写在 FriendBusyness 类里。

1. FriendBusyness 类

FriendBusyness 类的各方法功能说明如下。

（1）public boolean addFriend(Friend f,String userName)：添加朋友信息。

（2）public ArrayList<Friend> getAllFriends(String userName)：根据用户名查询该用户的朋友信息。

（3）public Friend getAFriend(String id)：根据朋友的 id 查询该朋友的信息。

（4）public void deleteFriend(String id[])：根据朋友的 id 删除朋友信息，删除一个或多个朋友。

（5）public boolean upadateFriend(Friend f)：更新朋友信息。

FriendBusyness 类的代码如下：

FriendBusyness. java

```java
package busyness;
import java.sql.Connection;
import java.sql.PreparedStatement;
import java.sql.ResultSet;
import java.sql.SQLException;
import java.util.ArrayList;
import common.DBConnection;
import entity.Friend;
public class FriendBusyness {
    //实现添加朋友信息
    public boolean addFriend(Friend f,String userName){
        boolean b=false;
        Connection con=DBConnection.getOneCon();
        PreparedStatement ps=null;
        try {
            ps=con.prepareStatement("insert into friendinfo values(?,?,to_date
            (?,'YYYY-MM-DD'),?,?,?,?,?)");
            ps.setString(1, f.getId());
            ps.setString(2, f.getName());
            ps.setString(3, f.getBirthday());             //日期类型
            ps.setString(4, f.getTelephone());
            ps.setString(5, f.getEmail());
            ps.setString(6, f.getAddress());
            ps.setString(7, f.getRelation());
            ps.setString(8, userName);
            int i=ps.executeUpdate();
            if(i>0)
                b=true;
        } catch (SQLException e) {
            //TODO Auto-generated catch block
            e.printStackTrace();
        }finally{
            DBConnection.close(ps);
            DBConnection.close(con);
        }
```

```java
        return b;
    }
    //查询朋友信息
    public ArrayList<Friend>getAllFriends(String userName){
        ArrayList<Friend>  al=new ArrayList<Friend>();
        Connection con=DBConnection.getOneCon();
        PreparedStatement ps=null;
        ResultSet rs=null;
        try {
            ps=con.prepareStatement("select * from friendinfo where userName=?");
            ps.setString(1, userName);
            rs=ps.executeQuery();
            while(rs.next()){
                Friend f=new Friend();
                f.setId(rs.getString(1));
                f.setName(rs.getString(2));
                f.setBirthday(rs.getString(3).substring(0, 10));  //取出 YYYY-MM-DD
                f.setTelephone(rs.getString(4));
                f.setEmail(rs.getString(5));
                f.setAddress(rs.getString(6));
                f.setRelation(rs.getString(7));
                al.add(f);
            }
        } catch (SQLException e) {
            //TODO Auto-generated catch block
            e.printStackTrace();
        }finally{
            DBConnection.close(rs);
            DBConnection.close(ps);
            DBConnection.close(con);
        }
        return al;
    }

    //查询一个朋友信息
    public Friend getAFriend(String id){
        Connection con=DBConnection.getOneCon();
        PreparedStatement ps=null;
        ResultSet rs=null;
        Friend f=new Friend();
        try {
            ps=con.prepareStatement("select * from friendinfo where id=?");
            ps.setString(1, id);
            rs=ps.executeQuery();
            if(rs.next()){
                f.setId(rs.getString(1));
                f.setName(rs.getString(2));
                f.setBirthday(rs.getString(3).substring(0, 10));
                                                        //取出 YYYY-MM-DD
                f.setTelephone(rs.getString(4));
                f.setEmail(rs.getString(5));
```

```java
            f.setAddress(rs.getString(6));
            f.setRelation(rs.getString(7));
        }
    } catch (SQLException e) {
        //TODO Auto-generated catch block
        e.printStackTrace();
    }finally{
        DBConnection.close(rs);
        DBConnection.close(ps);
        DBConnection.close(con);
    }
    return f;
}
//删除朋友信息
public void deleteFriend(String id[]){
    Connection con=DBConnection.getOneCon();
    PreparedStatement ps=null;
    try {
        ps=con.prepareStatement("delete from friendinfo where id=?");
        for(int i=0;i<id.length;i++){
            ps.setString(1, id[i]);
            ps.executeUpdate();
        }
    } catch (SQLException e) {
        //TODO Auto-generated catch block
        e.printStackTrace();
    }finally{
        DBConnection.close(ps);
        DBConnection.close(con);
    }
}
//修改朋友信息
public boolean upadateFriend(Friend f){
    boolean b=false;
    Connection con=DBConnection.getOneCon();
    PreparedStatement ps=null;
    try {
        ps=con.prepareStatement("update friendinfo set name=?," +
                "birthday=to_date(?,'YYYY-MM-DD')," +
                "telephone=?," +
                "email=?," +
                "address=?," +
                "relation=? where id=?");
        ps.setString(1, f.getName());
        ps.setString(2, f.getBirthday());           //日期类型
        ps.setString(3, f.getTelephone());
        ps.setString(4, f.getEmail());
        ps.setString(5, f.getAddress());
        ps.setString(6, f.getRelation());
        ps.setString(7, f.getId());
        int i=ps.executeUpdate();
```

```
            if(i>0)
                b=true;
        } catch (SQLException e) {
            //TODO Auto-generated catch block
            e.printStackTrace();
        }finally{
            DBConnection.close(ps);
            DBConnection.close(con);
        }
        return b;
    }
}
```

2. UserBusyness 类

UserBusyness 类的各方法功能说明如下。

（1）public boolean nameIsExit(String name)：判断用户名是否存在。

（2）public boolean regist(User u)：注册用户信息。

（3）public boolean login(User u)：实现登录功能。

（4）public boolean upadatePassword(User u)：更新用户的密码。

UserBusyness 类的代码如下：

UserBusyness.java

```
package busyness;
import java.sql.*;
import common.DBConnection;
import entity.User;
public class UserBusyness {
    //判断用户名是否可用
    public boolean nameIsExit(String name){
        boolean b=true;
        Connection con=DBConnection.getOneCon();
        PreparedStatement ps=null;
        ResultSet rs=null;
        try {
            ps=con.prepareStatement("select * from usertable where userName=?");
            ps.setString(1, name);
            rs=ps.executeQuery();
            if(rs.next())
                b=false;
        } catch (SQLException e) {
            //TODO Auto-generated catch block
            e.printStackTrace();
        }finally{
            DBConnection.close(rs);
            DBConnection.close(ps);
            DBConnection.close(con);
        }
        return b;
    }
```

```java
//实现注册功能
public boolean regist(User u){
    boolean b=false;
    Connection con=DBConnection.getOneCon();
    PreparedStatement ps=null;
    try {
        ps=con.prepareStatement("insert into usertable values(?,?)");
        ps.setString(1, u.getUname());
        ps.setString(2, u.getUpass());
        int i=ps.executeUpdate();
        if(i>0)
            b=true;
    } catch (SQLException e) {
        //TODO Auto-generated catch block
        e.printStackTrace();
    }finally{
        DBConnection.close(ps);
        DBConnection.close(con);
    }
    return b;
}
//实现登录功能
public boolean login(User u){
    boolean b=false;
    Connection con=DBConnection.getOneCon();
    PreparedStatement ps=null;
    ResultSet rs=null;
    try {
        ps=con.prepareStatement("select * from usertable where userName=? and password=?");
        ps.setString(1, u.getUname());
        ps.setString(2, u.getUpass());
        rs=ps.executeQuery();
        if(rs.next())
            b=true;
    } catch (SQLException e) {
        //TODO Auto-generated catch block
        e.printStackTrace();
    }finally{
        DBConnection.close(rs);
        DBConnection.close(ps);
        DBConnection.close(con);
    }
    return b;
}
//实现修改密码功能
public boolean upadatePassword(User u){
    boolean b=false;
    Connection con=DBConnection.getOneCon();
    PreparedStatement ps=null;
    try {
```

```
            ps=con.prepareStatement("update  usertable set  password =?  where
            userName =?");
            ps.setString(1, u.getUpass());
            ps.setString(2, u.getUname());
            int i=ps.executeUpdate();
            if(i>0)
                b=true;
        } catch (SQLException e) {
            //TODO Auto-generated catch block
            e.printStackTrace();
        }finally{
            DBConnection.close(ps);
            DBConnection.close(con);
        }
        return b;
    }
}
```

10.5 系统实现

10.5.1 用户注册

当新用户注册时,该模块要求用户必须输入用户姓名和密码信息,否则不允许注册。注册成功的用户信息被存入数据库的 usertable 表中。

1. 视图(JSP 页面)

该模块中只有一个 JSP 页面:regist.jsp,该页面负责提供注册信息的输入界面。另外,在注册页面中使用了一个隐藏域 flag,控制器 servlet 根据 flag 的值,确定实现注册还是判断用户名是否存在。

regist.jsp

效果如图 10.4 所示。

图 10.4 注册页面

```
<%@page language="java" contentType="text/html; charset=
GBK" pageEncoding="GBK"%>
<%@page isELIgnored="false" %>
<%@taglib prefix="c" uri="http://java.sun.com/jsp/jstl/core"%>
<%
String path=request.getContextPath();
String basePath = request.getScheme () +"://" + request.getServerName () +":" +
request.getServerPort()+path+"/";
%>
<html>
<head>
<base href="<%=basePath%>">
<style type="text/css">
    .textSize{
        width: 100pt;
        height: 15pt
    }
```

```
</style>
<title>注册画面</title>
<script type="text/javascript">
    //输入姓名后,调用该方法,判断用户名是否可用
    function nameIsNull(){
        var name=document.registForm.uname.value;
        if(name==""){
            alert("请输入姓名!");
            document.registForm.uname.focus();
            return false;
        }
        document.registForm.flag.value="0";
        document.registForm.submit();
        return true;
    }
    //注册时检查输入项
    function allIsNull(){
        var name=document.registForm.uname.value;
        var pwd=document.registForm.upass.value;
        var repwd=document.registForm.reupass.value;
        if(name==""){
            alert("请输入姓名!");
            document.registForm.uname.focus();
            return false;
        }
        if(pwd==""){
            alert("请输入密码!");
            document.registForm.upass.focus();
            return false;
        }
        if(repwd==""){
            alert("请输入确认密码!");
            document.registForm.reupass.focus();
            return false;
        }
        if(pwd!=repwd){
            alert("2次密码不一致,请重新输入!");
            document.registForm.upass.value="";
            document.registForm.reupass.value="";
            document.registForm.upass.focus();
            return false;
        }
        document.registForm.flag.value="1";
        document.registForm.submit();
        return true;
    }
</script>
</head>
<body>
    <form action="registServlet" method="post" name="registForm">
        <input type="hidden" name="flag">
```

```html
<table 
border=1 
bgcolor="lightblue" 
align="center">
    <tr>
        <td>姓名:</td>
        <td>
            <input class="textSize" type="text" name="uname" maxlength=
            "20" value="${requestScope.userName}"onblur="nameIsNull()"/>
            <c:if test="${requestScope.isExit=='false'}">
                <font color=red size=5>×</font>
            </c:if>
            <c:if test="${requestScope.isExit=='true'}">
                <font color=green size=5>√</font>
            </c:if>
        </td>
    </tr>

    <tr>
        <td>密码:</td>
        <td><input class="textSize" type="password" maxlength="20"
        name="upass"/></td>
    </tr>

    <tr>
        <td>确认密码:</td>
        <td><input class="textSize" type="password" maxlength="20"
        name="reupass"/></td>
    </tr>

    <tr>
        <td colspan="2" align="center"><input type="button" value="注
        册" onclick="allIsNull()"/></td>
    </tr>
</table>
    </form>
</body>
</html>
```

2. 控制器(servlet)

该控制器 servlet 对象的名称是 registServlet(见 10.3.4 小节的 web.xml 配置文件)。控制器获取视图的请求后,将视图中的信息封装在实体模型 User(见 10.4.3 小节)中。如果获取的请求是检查用户名(flag 等于 0)是否可用,则调用业务模型 UserBusyness(见 10.4.4 小节)中的 nameIsExit 方法执行业务处理。不管用户名是否存在,都返回 regist.jsp 页面告诉用户。如果获取的请求是注册(flag 等于 1),则调用业务模型 UserBusyness 中的 regist 方法实现注册。注册成功回到 login.jsp 页面,注册失败则回到 regist.jsp 页面。

RegistServlet.java

```java
package servlet;
import java.io.IOException;
import java.io.PrintWriter;
import javax.servlet.RequestDispatcher;
import javax.servlet.ServletException;
import javax.servlet.http.HttpServlet;
import javax.servlet.http.HttpServletRequest;
import javax.servlet.http.HttpServletResponse;
import busyness.UserBusyness;
import entity.User;
public class RegistServlet extends HttpServlet {
    protected void doGet(HttpServletRequest request, HttpServletResponse response)
            throws ServletException, IOException {
        response.setContentType("text/html;charset=GBK");
        PrintWriter out=response.getWriter();
        //获得页面提交的信息
        String name=request.getParameter("uname");
        String pass=request.getParameter("upass");
        String flag=request.getParameter("flag");
                                        //flag是个隐藏域,判断是检查用户还是注册
        //创建实体模型
        User u=new User();
        u.setUname(name);
        u.setUpass(pass);
        //创建业务模型
        UserBusyness ub=new UserBusyness();
        //检查用户名是否可用
        if(flag!=null&&flag.equals("0")){
            //用户名可用
            if(ub.nameIsExit(name)){
                request.setAttribute("isExit","true");
            }else{                      //用户名不可用
                request.setAttribute("isExit","false");
            }
            //不管用户名可不可用都返回页面继续注册
            request.setAttribute("userName",name);
            //回到注册画面
            RequestDispatcher dis=request.getRequestDispatcher("regist.jsp");
            dis.forward(request, response);
        }
        //实现注册功能
        if(flag!=null&&flag.equals("1")){
            if(ub.regist(u)){
                //注册成功回到登录画面
                out.print("注册成功,3秒后去登录!");
                response.setHeader("refresh", "3;url=login.jsp");
            }else{
                //失败回到注册画面
                out.print("注册失败,请查查原因,3秒后继续注册!");
```

```
            response.setHeader("refresh", "3;url=regist.jsp");
        }
    }
}
    protected void doPost(HttpServletRequest request, HttpServletResponse response)
            throws ServletException, IOException {
        doGet(request,response);
    }
}
```

10.5.2 用户登录

用户输入自己的用户的姓名和密码后,系统将对用户的姓名和密码进行验证。如果用户的姓名和密码同时正确,则成功登录,进入系统管理主页面(main.jsp);如果用户的姓名或密码有误,则回到登录页面继续登录。

1. 视图(JSP 页面)

login.jsp 页面提供登录信息输入的界面。效果如图 10.5 所示。

图 10.5 登录界面

login.jsp

```
<%@page language="java" contentType="text/html; charset=GBK" pageEncoding="GBK"%>
<%
String path=request.getContextPath();
String basePath = request.getScheme()+"://" + request.getServerName()+":" + request.getServerPort()+path+"/";
%>
<html>
<head>
<base href="<%=basePath%>">
<title>登录画面</title>
<style type="text/css">
    .textSize{
        width: 100pt;
        height: 15pt
    }
</style>
<script type="text/javascript">
//注册时检查输入项
    function allIsNull(){
        var name=document.loginForm.uname.value;
        var pwd=document.loginForm.upass.value;
        if(name==""){
            alert("请输入姓名!");
            document.loginForm.uname.focus();
            return false;
        }
        if(pwd==""){
            alert("请输入密码!");
            document.loginForm.upass.focus();
```

```
            return false;
        }
        document.loginForm.submit();
        return true;
    }
</script>
</head>
<body>
    <form action="login" method="post" name="loginForm">
        <table
        border=1
        bgcolor="lightblue"
        align="center">
            <tr>
                <td>姓名:</td>
                <td><input class="textSize" type="text" name="uname" maxlength
                ="20"/></td>
            </tr>

            <tr>
                <td>密码:</td>
                <td>< input class =" textSize" type =" password" name =" upass"
                maxlength="20"/></td>
            </tr>

            <tr>
                <td><input type="button" value="提交" onclick="allIsNull()"/>
                </td>
                <td>没有账号,请<a href="regist.jsp">注册</a>!</td>
            </tr>
        </table>
    </form>
</body>
</html>
```

2. 控制器(servlet)

该控制器 servlet 对象的名称是 loginServlet(见 10.3.4 小节的 web.xml 配置文件)。控制器获取视图的请求后,将视图中的信息封装在实体模型 User(见 10.4.3 小节)中,然后调用业务模型 UserBusyness(见 10.4.4 小节)中的 login 方法执行登录的业务处理。登录成功进入 main.jsp 页面,登录失败则返回到 login.jsp 页面。

LoginServlet.java

```
package servlet;
import java.io.IOException;
import java.io.PrintWriter;
import javax.servlet.RequestDispatcher;
import javax.servlet.ServletException;
import javax.servlet.http.HttpServlet;
import javax.servlet.http.HttpServletRequest;
```

```java
import javax.servlet.http.HttpServletResponse;
import javax.servlet.http.HttpSession;
import busyness.UserBusyness;
import entity.User;
public class LoginServlet extends HttpServlet {
    protected void doGet(HttpServletRequest request, HttpServletResponse response)
            throws ServletException, IOException {
        response.setContentType("text/html;charset=GBK");
        PrintWriter out=response.getWriter();
        //获得页面提交的信息
        String name=request.getParameter("uname");
        String pass=request.getParameter("upass");
        //创建实体模型 user
        User u=new User();
        u.setUname(name);
        u.setUpass(pass);
        //创建业务模型
        UserBusyness ub=new UserBusyness();
        //登录成功
        if(ub.login(u)){
            HttpSession session=request.getSession();
            //登录成功把用户名保存到 session 中
            session.setAttribute("user", u);
            RequestDispatcher dis=request.getRequestDispatcher("main.jsp");
            dis.forward(request, response);
        }else{//登录失败
            out.print("登录失败,查看用户名和密码是否错误？3秒钟后继续登录!");
            response.setHeader("refresh", "3;url=login.jsp");
        }
    }
    protected void doPost(HttpServletRequest request, HttpServletResponse response)
            throws ServletException, IOException {
        doGet(request,response);
    }
}
```

10.5.3 添加朋友信息

用户输入朋友姓名、生日、电话、E-mail、地址、关系等信息后,该模块首先检查输入的信息是否合法(比如,日期和 E-mail)。如果合法,则实现添加;如果不合法,则提示信息修改。

1. 视图(JSP 页面)

addFriend.jsp 页面实现添加朋友信息的输入界面。该页面中的朋友 ID 由主键产生器(见 10.4.2 小节中的 CreateID)自动产生。

addFriend.jsp

效果如图 10.6 所示。

```jsp
<%@page language="java" contentType="text/html; charset=GBK"
    pageEncoding="GBK"%>
<%@page import="common.*"%>
<%
```

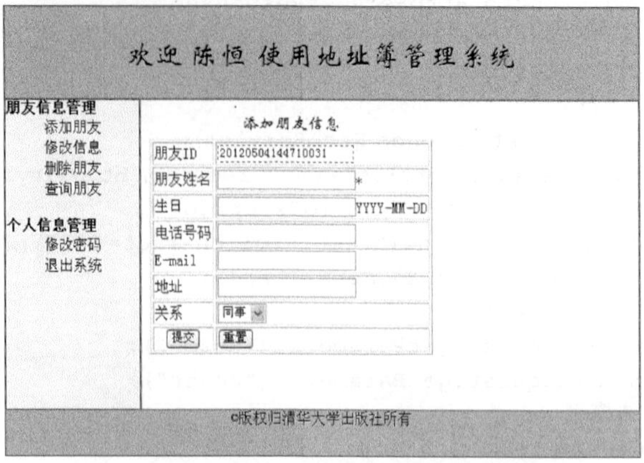

图 10.6 输入添加朋友的信息

```
        String path=request.getContextPath();
        String basePath=request.getScheme() +"://"
            +request.getServerName() +":" +request.getServerPort()
            +path +"/";
%>
<html>
<head>
<title>添加页面</title>
<script type="text/javascript" src="js/wpCalendar.js" charset="GBK"></script>
<script type="text/javascript">
//验证日期类型
function checkDate(da){
    varb=
/((^((1[8-9]\d{2})|([2-9]\d{3}))([-\/\._])(10|12|0?[13578])([-\/\._])(3[01]|
[12][0-9]|0?[1-9])$)|(^((1[8-9]\d{2})|([2-9]\d{3}))([-\/\._])(11|0?[469])
([-\/\._])(30|[12][0-9]|0?[1-9])$)|(^((1[8-9]\d{2})|([2-9]\d{3}))([-\/\._])
(0?2)([-\/\._])(2[0-8]|1[0-9]|0?[1-9])$)|(^([2468][048]00)([-\/\._])(0?2)
([-\/\._])(29)$)|(^([3579][26]00)([-\/\._])(0?2)([-\/\._])(29)$)|(^([1[89]
[0][48])([-\/\._])(0?2)([-\/\._])(29)$)|(^([2-9][0-9][0][48])([-\/\._])(0?2)
([-\/\._])(29)$)|(^([1[89][2468][048])([-\/\._])(0?2)([-\/\._])(29)$)|(^([2
-9][0-9][2468][048])([-\/\._])(0?2)([-\/\._])(29)$)|(^([1][89][13579][26])
([-\/\._])(0?2)([-\/\._])(29)$)|(^([2-9][0-9][13579][26])([-\/\._])(0?2)
([-\/\._])(29)$))/;
    if(!b.test(da)){
        return false;
    }else{
        return true;
    }
}
//验证邮箱地址
function checkEmail(da){
    var b=/^\w+((-\w+)|(\.\w+))*\@[A-Za-z0-9]+((\.|-)[A-Za-z0-9]+)*\.[A-Za-z0
-9]+$/;
    if(!b.test(da)){
        return false;
```

```
    }else{
        return true;
    }
}
//验证姓名是否输入、日期类型、邮箱格式
function nameisnull(){
    var name=document.addFriendForm.name.value;
    var day=document.addFriendForm.birthday.value;
    var email=document.addFriendForm.email.value;
    if(name==null||name.length==0){
        alert("请输入朋友姓名!");
        document.addFriendForm.name.focus();
        return false;
    }
    //验证日期类型
    if(day!=""&&day.length>0){
        if(!checkDate(day)){
            alert("日期格式不正确,请按照 YYYY-MM-DD 或 YYYY/MM/DD 或 YYYY-M-D 或
            YYYY/M/D 输入!");
            document.addFriendForm.birthday.value="";
            document.addFriendForm.birthday.focus();
            return false;
        }
    }
    //验证邮箱类型
    if(email!=""&&email.length>0){
        if(!checkEmail(email)){
            alert("E-mail 格式输入不正确!");
            document.addFriendForm.email.value="";
            document.addFriendForm.email.focus();
            return false;
        }
    }
    document.addFriendForm.submit();
    return true;
}
</script>
</head>
<body>
    <form action="addFriend" method="post" name="addFriendForm">
        <table border=1>
            <caption>
                <font size=4 face=华文新魏>添加朋友信息</font>
            </caption>
            <tr>
                <td>朋友 ID</td>
                <td><input type="text" readonly
                    style="border-width: 1pt; border-style: dashed; border-
                    color: red"
                    name="id" value='<%=CreateID.getStringID()%>'></td>
            </tr>
```

```html
<tr>
    <td>朋友姓名</td>
    <td><input type="text" name="name" maxlength="20" /><font color="red">*</font>
    </td>
</tr>

<tr>
    <td>生日</td>
    <td><input type="text" name="birthday" maxlength="10" onfocus="showCalendar(this)"  />YYYY-MM-DD</td>
</tr>

<tr>
    <td>电话号码</td>
    <td><input type="text" name="telephone" maxlength="20" />
    </td>
</tr>

<tr>
    <td>E-mail</td>
    <td><input type="text" name="email" maxlength="20" />
    </td>
</tr>

<tr>
    <td>地址</td>
    <td><input type="text" name="address" maxlength="50" />
    </td>
</tr>

<tr>
    <td>关系</td>
    <td><select name="relation">
        <option value="同事" />同事
        <option value="同学" />同学
        <option value="战友" />战友
        <option value="老乡" />老乡
        <option value="亲戚" />亲戚
        <option value="家人" />家人
        <option value="密友" />密友
        <option value="其他" />其他
    </select></td>
</tr>

<tr>
    <td align="center"><input type="button" onclick="nameisnull()" value="提交" />
    </td>
    <td align="left"><input type="reset" value="重置" />
```

```
            </td>
        </tr>
    </table>
</form>
</body>
</html>
```

wpCalendar.js

使用 JavaScript 在 JSP 页面中实现万年历的功能代码在 wpCalendar.js 文件中。由于代码繁多，在这里不再叙述，请参考本书的源代码 ch10。

2. 控制器(servlet)

该控制器 servlet 对象的名称是 addFriendServlet(见 10.3.4 小节的 web.xml 配置文件)。控制器获取视图的请求后，将视图中的信息封装在实体模型 Friend(见 10.4.3 小节)中，然后调用业务模型 FriendBusyness(见 10.4.4 小节)中的 addFriend 方法执行添加的业务处理。添加成功进入 queryFriend.jsp 页面，添加失败则回到 addFriend.jsp 页面。

AddFriendServlet.java

```java
package servlet;
import java.io.IOException;
import java.io.PrintWriter;
import javax.servlet.RequestDispatcher;
import javax.servlet.ServletException;
import javax.servlet.http.HttpServlet;
import javax.servlet.http.HttpServletRequest;
import javax.servlet.http.HttpServletResponse;
import javax.servlet.http.HttpSession;
import busyness.FriendBusyness;
import entity.*;
public class AddFriendServlet extends HttpServlet {
    protected void doGet(HttpServletRequest request, HttpServletResponse response)
            throws ServletException, IOException {
        //获得页面提交的信息
        String id=request.getParameter("id");
        String name=request.getParameter("name");
        String birthday=request.getParameter("birthday");
        String telephone=request.getParameter("telephone");
        String email=request.getParameter("email");
        String address=request.getParameter("address");
        String relation=request.getParameter("relation");
        //创建实体模型
        Friend f=new Friend();
        f.setId(id);
        f.setName(name);
        if(birthday==null||birthday.length()==0){
            birthday="1900-01-01";              //没有输入生日,设置默认值
        }
        f.setBirthday(birthday);
        if(telephone==null||telephone.length()==0){
            telephone="没有电话";                //没有输入电话时,设置默认值
        }
```

```java
            f.setTelephone(telephone);
            if(email==null||email.length()==0){
                email="没有email";                          //没有输入email时,设置默认值
            }
            f.setEmail(email);
            if(address==null||address.length()==0){
                address="没有地址";                         //没有输入地址时,设置默认值
            }
            f.setAddress(address);
            f.setRelation(relation);
            HttpSession session=request.getSession();
            //从session中获取用户名
            String userName=((User)session.getAttribute("user")).getUname();
            //创建业务模型
            FriendBusyness fb=new FriendBusyness();
            //添加成功回到查询页面
            if(fb.addFriend(f, userName)){
                RequestDispatcher dis=request.getRequestDispatcher("queryFriend");
                dis.forward(request, response);
            }else{
                //添加失败回到添加画面
                RequestDispatcher dis=request.getRequestDispatcher("addFriend.jsp");
                dis.forward(request, response);
            }
    }
    protected void doPost(HttpServletRequest request,HttpServletResponse response)
            throws ServletException, IOException {
        doGet(request,response);
    }
}
```

10.5.4 查询朋友信息

单击管理信息主页面中的"查询朋友"超链接,打开查询页面 queryFriend.jsp。该超链接的目标地址是个 servlet,servlet 对象的名称是 queryFriendServlet(见 10.3.4 小节的 web.xml 配置文件)。在该 servlet 控制器中,调用业务模型 FriendBusyness(见 10.4.4 小节)中的 getAllFriends 方法执行查询的业务处理,把查询结果显示在查询页面 queryFriend.jsp 中。

在页面 queryFriend.jsp 中单击"查看详情"超链接,打开详细信息页面 friendDetail.jsp。该超链接的目标地址是个 servlet,servlet 对象的名称是 friendDetailServlet(见 10.3.4 小节的 web.xml 配置文件)。在该 servlet 控制器中,调用业务模型 FriendBusyness (见 10.4.4 小节)中的 getAFriend 方法执行查询一个朋友信息的业务处理,根据 op 的值,把查询结果显示在详细信息页面 friendDetail.jsp 中。

1. 控制器(servlet)
QueryFriendServlet.java

```java
package servlet;
import java.io.IOException;
```

```java
import java.util.ArrayList;
import javax.servlet.RequestDispatcher;
import javax.servlet.ServletException;
import javax.servlet.http.HttpServlet;
import javax.servlet.http.HttpServletRequest;
import javax.servlet.http.HttpServletResponse;
import javax.servlet.http.HttpSession;
import busyness.FriendBusyness;
import entity.Friend;
import entity.User;
public class QueryFriendServlet extends HttpServlet {
    protected void doGet(HttpServletRequest request, HttpServletResponse response)
            throws ServletException, IOException {
        //创建业务模型
        FriendBusyness fb=new FriendBusyness();
        HttpSession session=request.getSession();
        //得到用户名
        String userName=((User)session.getAttribute("user")).getUname();
        //获得朋友信息
        ArrayList<Friend>al=fb.getAllFriends(userName);
        //把数组放到 request 里传给 JSP 页面
        request.setAttribute("friends", al);
        //获得链接参数
        String flag=request.getParameter("flag");
        //转发到删除页面
        if(flag!=null&&flag.equals("del")){
            RequestDispatcher dis=request.getRequestDispatcher("deleteFriend.
            jsp");
            dis.forward(request, response);
            return;
        }
        //转发到删除页面
        if(flag!=null&&flag.equals("update")){
            RequestDispatcher dis=request.getRequestDispatcher("modifyFriend.
            jsp");
            dis.forward(request, response);
            return;
        }
        //转发到查询页面
        RequestDispatcher dis = request. getRequestDispatcher (" queryFriend.
        jsp");
        dis.forward(request, response);

    }
    protected void doPost(HttpServletRequest request,HttpServletResponse response)
            throws ServletException, IOException {
        doGet(request,response);
    }
}
```

FriendDetailServlet. java

```java
package servlet;
```

```java
import java.io.IOException;
import javax.servlet.RequestDispatcher;
import javax.servlet.ServletException;
import javax.servlet.http.HttpServlet;
import javax.servlet.http.HttpServletRequest;
import javax.servlet.http.HttpServletResponse;
import busyness.FriendBusyness;
import entity.Friend;
public class FriendDetailServlet extends HttpServlet {
    protected void doGet(HttpServletRequest request, HttpServletResponse response)
            throws ServletException, IOException {
        //创建业务模型
        FriendBusyness fb=new FriendBusyness();
        //获得链接的参数
        String id=request.getParameter("friendid");
        String op=request.getParameter("op");

        //获得朋友信息
        Friend f=fb.getAFriend(id);
        request.setAttribute("friend", f);
        //查看详细信息
        if(op!=null&&op.equals("detail")){
            RequestDispatcher dis=request.getRequestDispatcher("friendDetail.jsp");
            dis.forward(request, response);
        }
        //查看修改的信息
        if(op!=null&&op.equals("update")){
            RequestDispatcher dis=request.getRequestDispatcher("friendupdate.jsp");
            dis.forward(request, response);
        }

    }
    protected void doPost(HttpServletRequest request,HttpServletResponse response)
            throws ServletException, IOException {
        doGet(request,response);
    }
}
```

2. 视图(JSP 页面)

queryFriend.jsp

效果如图 10.7 所示。

```
<%@page language="java" contentType="text/html; charset=GBK" pageEncoding="GBK"%>
<%@page isELIgnored="false" %>
<%@taglib prefix="c" uri="http://java.sun.com/jsp/jstl/core"%>
<%
String path=request.getContextPath();
```

图 10.7 朋友信息显示页面

```
String basePath = request. getScheme () +"://" + request. getServerName()+":"+
request.getServerPort()+path+"/";
%>
<html>
    <head>
     <base href="<%=basePath%>">
        <title>查询页面</title>
    </head>
    <body>
    <table border="1">
            <tr bgcolor="LightGreen">
                <th>朋友 ID</th>
                <th>朋友姓名</th>
                <th>关系</th>
                <th>查看详情</th>
            </tr>
    <c:forEach var="friend" items="${requestScope.friends}">
        <tr>
            <td>${friend.id}</td>
            <td>${friend.name}</td>
            <td>${friend.relation}</td>
            <td>< a href="friendDetail? op = detail&&friendid = ${friend.id}"
            target="_blank" style="text-decoration:none">查看详情</a></td>
        </tr>
    </c:forEach>
    </table>
    </body>
</html>
```

friendDetail. jsp

效果如图 10.8 所示。

图 10.8 详细信息页面

10.5.5 修改朋友信息

单击管理信息主页面中的"修改朋友"超链接,打开修改查询页面 modifyFriend.jsp。该超链接的目标地址是个 servlet,servlet 对象的名称是 queryFriendServlet(见 10.3.4 小节的 web.xml 配置文件)。在该 servlet 控制器(见 10.5.4 小节的控制器)中,根据 flag 的值,把查询结果显示在修改查询页面 modifyFriend.jsp 中。

单击 modifyFriend.jsp 页面中的"修改"超链接打开修改朋友信息页面 friendupdate.

jsp。该超链接的目标地址是个 servlet，servlet 对象的名称是 friendDetailServlet(见 10.3.4 小节的 web.xml 配置文件)。在该 servlet 控制器(见 10.5.4 小节的控制器)中，调用业务模型 FriendBusyness(见 10.4.4 小节)中的 getAFriend 方法执行查询一个朋友信息的业务处理。根据 op 的值，把查询结果显示在页面 friendupdate.jsp 中，然后进行朋友信息的修改。

输入要修改的信息后，单击"修改"按钮，把要修改的朋友信息提交给控制器 servlet，servlet 对象的名称是 updateFriendServlet(见 10.3.4 小节的 web.xml 配置文件)。在该 servlet 中将视图中的信息封装在实体模型 Friend(见 10.4.3 小节)中，然后调用业务模型 FriendBusyness(见 10.4.4 小节)中的 upadateFriend 方法执行修改的业务处理。成功修改进入 queryFriend.jsp 页面，修改失败回到 friendupdate.jsp 页面。

1. 视图(JSP 页面)

在该模块中视图共有两个：modifyFriend.jsp 和 friendupdate.jsp。modifyFriend.jsp 页面显示用户可以修改的朋友信息。friendupdate.jsp 页面提供修改信息输入界面。

modifyFriend.jsp

效果如图 10.9 所示。

图 10.9 可以修改的朋友信息

```jsp
<%@page language="java" contentType="text/html; charset=GBK"pageEncoding="GBK"%>
<%@page isELIgnored="false" %>
<%@taglib prefix="c" uri="http://java.sun.com/jsp/jstl/core"%>
<%
    String path=request.getContextPath();
    String basePath=request.getScheme() +"://"
        +request.getServerName() +":" +request.getServerPort()
        +path +"/";
%>
<html>
<head>
<base href="<%=basePath%>">
<title>修改页面</title>
</head>
<body>
        <table border="1">
            <tr bgcolor="LightGreen">
                <th>朋友 ID</th>
                <th>朋友姓名</th>
                <th>关系</th>
                <th>查看详情</th>
```

```
            </tr>
            <c:forEach var="friend" items="${requestScope.friends}">
               <tr>
                  <td>${friend.id}</td>
                  <td>${friend.name}</td>
                  <td>${friend.relation}</td>
                  <td align=" center " > < a href =" friendDetail? op =
                  update&&friendid = ${friend.id}" style =" text - decoration:
                  none">修改</a>
                  </td>
               </tr>
            </c:forEach>
         </table>
      </form>
</body>
</html>
```

friendupdate.jsp

效果如图 10.10 所示。

图 10.10 修改信息输入界面

```
<%@page language="java" contentType="text/html; charset=GBK" pageEncoding=
"GBK"%>
<%@page isELIgnored="false" %>
<%@taglib prefix="c" uri="http://java.sun.com/jsp/jstl/core"%>
<%
String path=request.getContextPath();
String basePath = request.getScheme()+"://" + request.getServerName()+":" +
request.getServerPort()+path+"/";
%>
<html>
<head>
<base href="<%=basePath%>">
<title>修改画面</title>
</head>
<script type="text/javascript" src="js/wpCalendar.js" charset="GBK"></script>
<script type="text/javascript">
//验证日期类型
function checkDate(da){
     var b=
```

```
/(((^((1[8-9]\d{2})|([2-9]\d{3}))([-\/\._])(10|12|0?[13578])([-\/\._])(3[01]|
[12][0-9]|0?[1-9])$)|(^((1[8-9]\d{2})|([2-9]\d{3}))([-\/\._])(11|0?[469])
([-\/\._])(30|[12][0-9]|0?[1-9])$)|(^((1[8-9]\d{2})|([2-9]\d{3}))([-\/\._])
(0?2)([-\/\._])(2[0-8]|1[0-9]|0?[1-9])$)|(^([2468][048]00)([-\/\._])(0?2)
([-\/\._])(29)$)|(^([3579][26]00)([-\/\._])(0?2)([-\/\._])(29)$)|(^([1][89]
[0][48])([-\/\._])(0?2)([-\/\._])(29)$)|(^([2-9][0-9][0][48])([-\/\._])(0?2)
([-\/\._])(29)$)|(^([1][89][2468][048])([-\/\._])(0?2)([-\/\._])(29)$)|(^([2
-9][0-9][2468][048])([-\/\._])(0?2)([-\/\._])(29)$)|(^([1][89][13579][26])
([-\/\._])(0?2)([-\/\._])(29)$)|(^([2-9][0-9][13579][26])([-\/\._])(0?2)
([-\/\._])(29)$))/;
    if(!b.test(da)){
        return false;
    }else{
        return true;
    }
}
//验证邮箱地址
function checkEmail(da){
    var b=/^\w+((-\w+)|(\.\w+))*\@[A-Za-z0-9]+((\.|-)[A-Za-z0-9]+)*\.[A-Za
-z0-9]+$/;
    if(!b.test(da)){
        return false;
    }else{
        return true;
    }
}
//验证姓名是否输入、日期类型、邮箱格式
function nameisnull(){
    var name=document.updateForm.name.value;
    var day=document.updateForm.birthday.value;
    var email=document.updateForm.email.value;
    if(name==null||name.length==0){
        alert("请输入朋友姓名!");
        document.updateForm.name.focus();
        return false;
    }
    //验证日期类型
    if(day!=""&&day.length>0){
        if(!checkDate(day)){
            alert("日期格式不正确,请按照 YYYY-MM-DD 或 YYYY/MM/DD 或 YYYY-M-D 或
            YYYY/M/D 输入!");
            document.updateForm.birthday.value="";
            document.updateForm.birthday.focus();
            return false;
        }
    }
    //验证邮箱类型
    if(email!=""&&email.length>0){
        if(!checkEmail(email)){
            alert("E-mail 格式输入不正确!");
            document.updateForm.email.value="";
```

```
            document.updateForm.email.focus();
            return false;
        }
    }
    document.updateForm.submit();
    return true;
}
</script>
<body bgcolor="LightCyan">
<form action="updateFriend" method="post" name="updateForm">
<table border=1>
        <caption><font size=4 face=华文新魏>朋友信息修改</font></caption>
        <tr>
            <td>朋友 ID</td>
            <td><input type="text" readonly
                style="border-width:1pt;border-style:dashed;border-color:red"
                name="id" value="${requestScope.friend.id}"/></td>
        </tr>

        <tr>
            <td>朋友姓名</td>
            <td>< input  type =" text"  name =" name"  maxlength =" 20"  value ="
            ${requestScope.friend.name}"/><font color="red"> * </font></td>
        </tr>

        <tr>
            <td>生日</td>
            <td>< input type =" text" name =" birthday" maxlength =" 10" value ="
            ${requestScope.friend.birthday}" onfocus="showCalendar(this)" />
            YYYY-MM-DD</td>
        </tr>

        <tr>
            <td>电话号码</td>
            <td><input type="text" name="telephone" maxlength="20"
        value="${requestScope.friend.telephone}"/></td>
        </tr>

        <tr>
            <td>E-mail</td>
            <td><input type="text" name="email" maxlength="20"
     value="${requestScope.friend.email}"/></td>
        </tr>

        <tr>
            <td>地址</td>
            <td><input type="text" name="address" maxlength="50"
       value="${requestScope.friend.address}"/></td>
        </tr>

        <tr>
```

```
<td>关系</td>
<td>
    <select name="relation" >
    <c:if test="${requestScope.friend.relation=='同事'}">
        <option value="同事" selected/>同事
        <option value="同学"/>同学
        <option value="战友"/>战友
        <option value="老乡"/>老乡
        <option value="亲戚"/>亲戚
        <option value="家人"/>家人
        <option value="密友"/>密友
        <option value="其他"/>其他
    </c:if>
    <c:if test="${requestScope.friend.relation=='同学'}">
        <option value="同事"/>同事
        <option value="同学" selected/>同学
        <option value="战友"/>战友
        <option value="老乡"/>老乡
        <option value="亲戚"/>亲戚
        <option value="家人"/>家人
        <option value="密友"/>密友
        <option value="其他"/>其他
    </c:if>

    <c:if test="${requestScope.friend.relation=='战友'}">
        <option value="同事"/>同事
        <option value="同学"/>同学
        <option value="战友" selected/>战友
        <option value="老乡"/>老乡
        <option value="亲戚"/>亲戚
        <option value="家人"/>家人
        <option value="密友"/>密友
        <option value="其他"/>其他
    </c:if>
    <c:if test="${requestScope.friend.relation=='老乡'}">
        <option value="同事"/>同事
        <option value="同学"/>同学
        <option value="战友"/>战友
        <option value="老乡" selected/>老乡
        <option value="亲戚"/>亲戚
        <option value="家人"/>家人
        <option value="密友"/>密友
        <option value="其他"/>其他
    </c:if>
    <c:if test="${requestScope.friend.relation=='亲戚'}">
        <option value="同事"/>同事
        <option value="同学" />同学
        <option value="战友"/>战友
        <option value="老乡"/>老乡
        <option value="亲戚" selected/>亲戚
        <option value="家人"/>家人
```

```
                <option value="密友"/>密友
                <option value="其他"/>其他
            </c:if>
            <c:if test="${requestScope.friend.relation=='家人'}">
                <option value="同事"/>同事
                <option value="同学" />同学
                <option value="战友"/>战友
                <option value="老乡"/>老乡
                <option value="亲戚"/>亲戚
                <option value="家人" selected />家人
                <option value="密友"/>密友
                <option value="其他"/>其他
            </c:if>
            <c:if test="${requestScope.friend.relation=='密友'}">
                <option value="同事"/>同事
                <option value="同学" />同学
                <option value="战友"/>战友
                <option value="老乡"/>老乡
                <option value="亲戚"/>亲戚
                <option value="家人"/>家人
                <option value="密友" selected />密友
                <option value="其他"/>其他
            </c:if>
            <c:if test="${requestScope.friend.relation=='其他'}">
                <option value="同事"/>同事
                <option value="同学" />同学
                <option value="战友"/>战友
                <option value="老乡"/>老乡
                <option value="亲戚"/>亲戚
                <option value="家人"/>家人
                <option value="密友"/>密友
                <option value="其他" selected />其他
            </c:if>
            </select>
        </td>
    </tr>
    <tr>
    <td colspan =" 2" align =" center " > < input type =" button " onclick ="nameisnull()"
    value="修改"/></td>
    </tr>
    </table>
</form>
</body>
</html>
```

2. 控制器(servlet)
UpdateFriendServlet.java

```
package servlet;
```

```java
import java.io.IOException;
import javax.servlet.RequestDispatcher;
import javax.servlet.ServletException;
import javax.servlet.http.HttpServlet;
import javax.servlet.http.HttpServletRequest;
import javax.servlet.http.HttpServletResponse;
import entity.Friend;
import busyness.FriendBusyness;
public class UpdateFriendServlet extends HttpServlet {
    protected void doGet(HttpServletRequest request, HttpServletResponse response)
            throws ServletException, IOException {
        //获得页面提交的信息
        String id=request.getParameter("id");
        String name=request.getParameter("name");
        String birthday=request.getParameter("birthday");
        String telephone=request.getParameter("telephone");
        String email=request.getParameter("email");
        String address=request.getParameter("address");
        String relation=request.getParameter("relation");
        //创建实体模型
        Friend f=new Friend();
        f.setId(id);
        f.setName(name);
        if(birthday==null||birthday.length()==0){
            birthday="1900-01-01";                    //没有输入生日,设置默认值
        }
        f.setBirthday(birthday);
        if(telephone==null||telephone.length()==0){
            telephone="没有电话";                      //没有输入电话时,设置默认值
        }
        f.setTelephone(telephone);
        if(email==null||email.length()==0){
            email="没有 email";                        //没有输入 email 时,设置默认值
        }
        f.setEmail(email);
        if(address==null||address.length()==0){
            address="没有地址";                        //没有输入地址时,设置默认值
        }
        f.setAddress(address);
        f.setRelation(relation);
        //创建业务模型
        FriendBusyness fb=new FriendBusyness();
        if(fb.upadateFriend(f)){
            RequestDispatcher dis=request.getRequestDispatcher("queryFriend");
            dis.forward(request, response);
        }else{
            //修改失败回到修改画面
            request.setAttribute("friend", f);        //返回输入的值
            RequestDispatcher dis=request.getRequestDispatcher("freindupdate.jsp");
            dis.forward(request, response);
```

```
            }
        }
    protected void doPost(HttpServletRequest request, HttpServletResponse response)
            throws ServletException, IOException {
        doGet(request,response);
    }
}
```

10.5.6 删除朋友信息

单击管理信息主页面中的"删除朋友"超链接，打开删除查询页面 deleteFriend.jsp。该超链接的目标地址是个 servlet，servlet 对象的名称是 queryFriendServlet（见 10.3.4 小节的 web.xml 配置文件）。在该 servlet 控制器（见 10.5.4 小节的控制器）中，根据 flag 的值，把查询结果显示在删除查询页面 deleteFriend.jsp 中。

在 deleteFriend.jsp 页面选中要删除的朋友的复选框，单击"删除"按钮，把要删除的朋友的 ID 提交给控制器 servlet，servlet 对象的名称是 deleteServlet（见 10.3.4 小节的 web.xml 配置文件）。在该 servlet 中调用业务模型 FriendBusyness（见 10.4.4 小节）中的 deleteFriend 方法执行删除的业务处理。成功删除后进入删除查询页面 deleteFriend.jsp。

1. 视图(JSP 页面)

deleteFriend.jsp

效果如图 10.11 所示。

图 10.11 可以删除的朋友信息

```
<%@page language="java" contentType="text/html; charset=GBK" pageEncoding="GBK"%>
<%@page isELIgnored="false" %>
<%@taglib prefix="c" uri="http://java.sun.com/jsp/jstl/core"%>
<%
String path=request.getContextPath();
String basePath = request.getScheme()+"://"+request.getServerName()+":"+request.getServerPort()+path+"/";
%>
<html>
    <head>
        <base href="<%=basePath%>">
        <title>删除页面</title>
        <script type="text/javascript">
            function confirmDelete(){
                var n=document.deleteForm.deleteid.length;
```

```
            var count=0;
            for(var i=0;i<n;i++){
                if(!document.deleteForm.deleteid[i].checked){
                    count++;
                }else{
                    break;
                }
            }
            if(n>1){
                //所有的朋友都没有选择
                if(count==n){
                    alert("请选择绝交的朋友!");
                    count=0;
                    return false;

                }
            }else{
                //就一个朋友并且还没有选择
                if(!document.deleteForm.deleteid.checked){
                        alert("请选择绝交的朋友!");
                        return false;
                    }
                }

            if(window.confirm("真的删除吗? really?")){
                document.deleteForm.submit();
                return true;
            }
            return false;
        }
</script>
</head>
<body>
    <form action="delete" name="deleteForm" method="post">
<table border="1">
        <tr bgcolor="LightGreen">
            <th>朋友 ID</th>
            <th>朋友姓名</th>
            <th>关系</th>
            <th>查看详情</th>
        </tr>
    <c:forEach var="friend" items="${requestScope.friends}">
    <tr>
            <td>< input type="checkbox" name="deleteid" value="${friend.id}"/>
        ${friend.id}</td>
            <td>${friend.name}</td>
            <td>${friend.relation}</td>
            <td> < a href=" friendDetail? op=detail&&friendid = ${friend.id}"
            target="_blank"
            style="text-decoration:none">查看详情</a></td>
    </tr>
```

```
            </c:forEach>
            <tr>
                <td colspan="4" align="center"><input type="button" value="删除"
                  onclick="confirmDelete()"/></td>
            </tr>
        </table>
        </form>
    </body>
</html>
```

2. 控制器(servlet)

DeleteServlet.java

```
package servlet;
import java.io.IOException;
import javax.servlet.RequestDispatcher;
import javax.servlet.ServletException;
import javax.servlet.http.HttpServlet;
import javax.servlet.http.HttpServletRequest;
import javax.servlet.http.HttpServletResponse;
import busyness.FriendBusyness;
public class DeleteServlet extends HttpServlet {
    protected void doGet(HttpServletRequest request, HttpServletResponse response)
            throws ServletException, IOException {
        //创建业务模型
        FriendBusyness fb=new FriendBusyness();
        //获得要删除的id
        String id[]=request.getParameterValues("deleteid");
        //删除朋友
        fb.deleteFriend(id);
        RequestDispatcher dis=request.getRequestDispatcher("queryFriend? flag
        =del");
        dis.forward(request, response);
    }
    protected void doPost(HttpServletRequest request,HttpServletResponse response)
            throws ServletException, IOException {
        doGet(request,response);
    }
}
```

10.5.7 修改密码

单击管理信息主页面中的"修改密码"超链接,打开密码修改页面 upadatepassword.jsp。在该页面中用户输入新密码,单击"修改"按钮。将密码信息提供给控制器 servlet,servlet 对象的名称是 updatePasswordServlet(见 10.3.4 小节的 web.xml 配置文件)。控制器获取视图的请求后,将视图中的信息封装在实体模型 User(见 10.4.3 小节)中,然后调用业务模型 UserBusyness(见 10.4.4 小节)中的 upadatePassword 方法执行修改密码的业务。

1. 视图(JSP 页面)
upadatepassword.jsp

效果如图 10.12 所示。

图 10.12 密码修改页面

```jsp
<%@page language="java" contentType="text/html; charset=GBK" pageEncoding="GBK"%>
<%@page isELIgnored="false" %>
<%@taglib prefix="c" uri="http://java.sun.com/jsp/jstl/core"%>
<%
String path=request.getContextPath();
String basePath = request.getScheme() +"://" + request.getServerName()+":"+request.getServerPort()+path+"/";
%>
<html>
<head>
<base href="<%=basePath%>">
<style type="text/css">
    .textSize{
        width: 100pt;
        height: 15pt
    }
</style>
<title>修改密码</title>
<script type="text/javascript">
    //注册时检查输入项
    function allIsNull(){
        var pwd=document.updatePasswordForm.upass.value;
        var repwd=document.updatePasswordForm.reupass.value;
        if(pwd==""){
            alert("请输入确认密码!");
            document.updatePasswordForm.upass.focus();
            return false;
        }
        if(repwd==""){
            alert("请输入确认密码!");
            document.updatePasswordForm.reupass.focus();
            return false;
        }
        if(pwd!=repwd){
            alert("2次密码不一致,请重新输入!");
            document.updatePasswordForm.upass.value="";
            document.updatePasswordForm.reupass.value="";
            document.updatePasswordForm.upass.focus();
            return false;
        }
        document.updatePasswordForm.submit();
        return true;
    }
</script>
</head>
```

```
<body>
    <form action="updatePassword" method="post" name="updatePasswordForm">
        <table
        border=1
        bgcolor="lightblue"
        align="center">
            <tr>
                <td>姓名:</td>
                <td>
                    ${sessionScope.user.uname }
                </td>
            </tr>

            <tr>
                <td>新密码:</td>
                <td><input class="textSize" type="password" name="upass"
                value="${sessionScope.user.upass }"/></td>
            </tr>

            <tr>
                <td>确认密码:</td>
                <td><input class="textSize" type="password" maxlength="20"
                name="reupass"/></td>
            </tr>


            <tr>
                <td colspan="2" align="center"><input type="button" value="修改"
                onclick="allIsNull()"/></td>
            </tr>
        </table>
    </form>
</body>
</html>
```

2. 控制器(servlet)
UpdatePasswordServlet.java

```
package servlet;
import java.io.IOException;
import java.io.PrintWriter;
import javax.servlet.ServletException;
import javax.servlet.http.HttpServlet;
import javax.servlet.http.HttpServletRequest;
import javax.servlet.http.HttpServletResponse;
import javax.servlet.http.HttpSession;
import busyness.UserBusyness;
import entity.User;
public class UpdatePasswordServlet extends HttpServlet {
    protected void doGet(HttpServletRequest request, HttpServletResponse response)
```

```
        throws ServletException, IOException {
    response.setContentType("text/html;charset=GBK");
    PrintWriter out=response.getWriter();
    //获得页面提交的信息
    String pass=request.getParameter("upass");
    HttpSession session=request.getSession();
    String name=((User)session.getAttribute("user")).getUname();
    //创建实体模型user
    User u=new User();
    u.setUname(name);
    u.setUpass(pass);
    //创建业务模型
    UserBusyness ub=new UserBusyness();
    //实现修改功能
    if(ub.upadatePassword(u)){
        //注册成功回到登录画面
        out.print("修改成功,3秒后去登录!");
        response.setHeader("refresh", "3;url=login.jsp");
    }else{
        //失败回到注册画面
        out.print("修改失败,请查查原因,3秒后继续修改!");
        response.setHeader("refresh", "3;url=upadatepassword.jsp");
    }
}
protected void doPost(HttpServletRequest request, HttpServletResponse response)
        throws ServletException, IOException {
    doGet(request,response);
}
}
```

10.5.8 退出系统

单击管理信息主页面中的"退出系统"超链接,该链接的目标地址是个控制器 servlet,servlet 对象的名称是 exitUserServlet(见 10.3.4 小节的 web.xml 配置文件)。在控制器中清空用户的会话 session。退出系统成功后,进入登录页面,重新登录。

控制器(servlet)

ExitUserServlet.java

```
package servlet;
import java.io.IOException;
import java.io.PrintWriter;
import javax.servlet.ServletException;
import javax.servlet.http.HttpServlet;
import javax.servlet.http.HttpServletRequest;
import javax.servlet.http.HttpServletResponse;
import javax.servlet.http.HttpSession;
public class ExitUserServlet extends HttpServlet {
    protected void doGet(HttpServletRequest request, HttpServletResponse response)
            throws ServletException, IOException {
```

```java
        response.setContentType("text/html;charset=GBK");
        PrintWriter out=response.getWriter();
        HttpSession session=request.getSession();
        session.removeAttribute("user");
        out.print("退出系统,3秒后重新登录!");
        response.setHeader("refresh", "3;url=login.jsp");
    }
    protected void doPost(HttpServletRequest request,HttpServletResponse response)
            throws ServletException, IOException {
        doGet(request,response);
    }
}
```